TALES OF ASHOK

When you are at your lowest ebb from illness or loss, and things are falling apart, you become strangely able to experience the world in new ways. 5,000 years ago a man called Ashok experienced exactly this when he heard a voice calling him to leave his place in the desert and go on a long journey. He was not told why or where but he travelled widely and, wherever he went, he met people who were confused and fearful and conflicted as wars and natural disasters overwhelmed them.

From encounters with kings and untouchables, yak herdsmen and masters of the financial universe, the rich, the poor, the disabled and the ordinary he drew wisdom and developed with them the wonder of new purpose and fresh beginnings.

These tales recount the epiphanies experienced by his Followers and the schools they set up to share the values and ways of being of the world they wanted to create.

The Tales are dedicated to
all who have the common
good of love in their hearts

by the same author – Leslie Dighton

Two Bunions and One Lottery Ticket
Walking on Sore Feet

The Little Books of the Common Good
Notes from Exile

by the same illustrator – Jennifer Williams

Across the Street Around the World
Creative Community Building
Common Threads Uncommon People

LESLIE DIGHTON

TALES OF ASHOK

illustrations by
Jennifer Williams

UMBRIA PRESS
talesofashok.com

First published in 2026 by Umbria Press
Umbria Press
London SW15 5DP
www.umbriapress.co.uk

talesofashok.com

Design by Louise Millar, London, UK

Printed in Poland by Totem
www.totem.com.pl

ISBN 978 1 910074 51 0

CONTENTS

TALES OF ASHOK

A fable for our present confused and conflicted times with magic at its heart

People the world over are confused and angry. They can see the big social systems aren't working anymore and are desperate for solutions. Nations compete for advantage and build walls to keep others out. Governments pursue growth above well-being and consumers happily consume. Entitlement overwhelms duty and accountability. Violence becomes the default option for resolving differences.

The visible signs are war, environmental destruction, climate turmoil and flows of migrants. Beneath the surface ordinary people can't find a community of shared purpose and increasingly opt for extreme answers. The beneficiaries are welfare claimants, arms suppliers and extremist politicians. The potential for apocalypse in our time grows ever greater and closer.

The Tales tell a story of society sleepwalking to destruction through the eyes of Ashok, a desert dweller, 5,000 years ago. He is called to go on a journey to find out why this is happening. Gifted with magical powers he travels through time and space to consult with the great civilisations, past and present. His findings offend those presently in power who take a terrible revenge. Ultimately, the earth collapses and imposes its own solution. In the vacuum of collapse, Ashok's Followers build a new world based on new values and ways of being through the Schools of the Common Good which spring up to rediscover hope, community, and co-operation.

Ashok becomes the Teacher of that New World.

ASHOK
the first
DUSTMAN

A long time ago in a faraway land, a man stumbled out of the desert on to a city on the banks of a mighty river. His name was Ashok. The river was the Euphrates. The city, which shimmered in the morning light, was called Ur, the capital of Sumeria and according to legend the birthplace of Abraham, the great Teacher.

Many nights before Ashok had been woken by a voice in the sky telling him to set forth on a long journey. He told his mother, Gaia, that the voice had spoken with such gentle authority he felt compelled to do as he was bidden. She looked into his eyes intensely for a long time as though seeing something she had never seen before.

She asked him no questions, only that he always wear the amulet she slipped around his neck and return to her when his journey was over. He had set out at the first sight of dawn, crossing endless horizons of sand dunes under a relentless sun by day and a freezing night sky.

Now, parched and exhausted, he gazed with wonder at the beauty and bustling city of Ur, and knew instinctively that it was to here, where the crossroads of the ancient world came together, that he had been called. To do what, he did not know.

BLACK SEA
EUROPE
CASPIAN
CHINA
TURKEY
SEA
INDIA
ASHOK'S JOURNEY
MEDITERRANEAN SEA
EUPHRATES
EGYPT
RED SEA
PERSIAN
ARABIA
GULF

Ashok had covered vast distances on foot through blistering heat and sandstorms by day and icy temperatures at night. He was physically exhausted and at the very limit of his endurance. What he wanted to do most of all was to find somewhere to lie down, rest and close his eyes.

But as he entered the city, astonished by its beauty and how busy and lively the streets were, his spirits rose and the weariness lifted from him.

He watched with awe at caravans arriving from countries far to the North, the South, the East and the West, camels and mules swaying under their heavy loads of textiles, ivory and exotic fruits. He decided to follow them to the caravanserai where they would unload and bed down, confident he would at least find a place to sleep with the animals.

The traders, whose shopfronts spilled out onto the pavements, offering spices and sweetmeats and household goods of every description, watched greedily as they passed through the central market place. "More business for us," they were thinking.

As he followed the caravans Ashok had to cover his face against the toxic smell coming from rubbish piled high in every street and alleyway.

The next day, rested but hungry, he wondered out loud to himself how he was going to cope in this strange place, where he knew no one. He could taste on his breath the bitter smell of camel after a night nestled for warmth against the side of a huge accommodating beast. But he preferred that to the smell of the rotting refuse which covered the streets.

His immediate need was to find work to pay for food and shelter. But above all he needed to find out why the voice in the sky had called him in such a commanding way to come to Ur.

Idly fingering the amulet, his constant companion hanging round his neck, he asked himself again, "Where do I start? What can I possibly offer this busy place which has already so much?" His thoughts instinctively turned to his mother, Gaia, who had pressed the amulet into his hand when he left, with the reassuring words that it would bring him comfort and security when needed. Good fortune had helped him through the desert to get this far. Perhaps once again he would be lucky and find work and shelter.

Encouraged by the contact he said farewell to his companion of the night, who bared her teeth in response and hissed what, despite the foulness of her breath, he took to be a token of affection.

Meanwhile, unknown to Ashok, in another part of the city the Elders were meeting that same day to consider the biggest problem facing Ur, which was so obvious it could no longer be ignored.

It was top of the list of citizens' complaints and highly visible through the windows of the Council Chamber itself. The 'it' that was causing so much concern was the problem of rubbish being dumped daily in the streets in ever increasing amounts, the smell of its decay hanging in the air and entering the homes of the rich and poor alike.

Angrily, the Elders had the windows closed in a vain effort to keep the problem at bay, but that didn't solve it. They were at their wits' end, but finding a solution was proving impossible. They also knew that the source of the problem was a good one for them and got them re-elected. Ur was growing fast and its citizens had never had it so good, consuming more and more 'stuff' brought in from distant lands to satisfy their ever growing appetite.

Its consequences, however, were not so good. More and more waste was being thrown out on the streets and left there to rot. The stench had become unbearable. People feared the plague in the poorer quarters would spread. The Elders knew something had to be done urgently, or the people would turn against them. But what to do?

When they heard rumours of a stranger looking for work they had a brainwave. They would set him the task of clearing the rubbish from the streets. They had no idea how he might do it but they would get credit for doing something and would have someone to blame if it failed.

Delighted with their idea, they quickly sent out messengers to find Ashok, who was wandering the streets searching despondently by this time for work at each of the shops and inns he passed. He couldn't believe his good fortune and readily accompanied the messenger to the Council Chamber where he was offered and accepted the job with delight.

And this is how Ashok became the first dustman of the modern world.

It didn't take long for his initial excitement to pass and to realise he had never actually encountered the problem of excess and waste before. In fact he had no experience of the modern world with surplus resources. Food in the desert had always been scarce, water treasured. Even the stones of the dates he ate were crushed and mixed with the goat swill, the fronds of the palms plaited to provide shelter. Nothing in his life had prepared him for an abundance of things which got thrown away.

And now, suddenly, he had the responsibility of cleaning up a whole city from its own excess.

He woke the next morning apprehensive but proud that he now had a very important job, and it was in his character to take it very seriously. He had no way of knowing, of course, he was the first official dustman ever to be appointed since the beginning of time. Nor that the problem of waste was going to be one of the big problems that the whole world, not just Ur, would have to address.

He just knew he had a big problem on his hands and it was getting bigger by the day. He could see with his own eyes the mountain of rubbish thrown out onto the streets was growing faster than he could possibly dispose of it.

There was no one he could turn to, no course he could sign up for. He just had to work it out for himself.

The obvious place to start was to talk to people he met putting out their rubbish onto the pavements to see if they could somehow be persuaded to reduce the quantity. Mostly they just looked at him and shrugged. "What do you expect us to do?" they asked. "We can't keep it indoors." Some got quite angry and told Ashok to mind his own business. "What right have you got to interfere with how I choose to live?"

Sometimes he had to back off smartly, fearing for his life.

It wasn't long before Ashok began to despair. The piles of rubbish he had to collect every day and somehow dispose of simply grew and grew, to such an extent he began to lose heart. Try as he might he just couldn't see any way out of the problem. For the first time in his life he felt absolutely alone and powerless.

In his darkest moments, he asked himself, "Perhaps I should just disappear back into the desert where life was simple." But throughout his life he had never shirked a challenge and the voice which still resonated in his head drove him on. He was ashamed that he could even think of giving up.

Although the response of the citizens he spoke to wasn't helpful, the problem was real enough and wouldn't go away of its own accord. It couldn't just be shrugged off. Somehow he had to find a solution.

To make matters worse the plague in the city was spreading and, fearful for their own safety, the citizens of Ur were loud in their criticism of the refuse that piled up in their streets and the acrid smell that filled the air.

"We don't want this in our backyard," was the message to Ashok. "Get it sorted."

Racking his brains in desperation, Ashok suddenly had an idea. He had no control over the amount of rubbish that was being put out. People didn't want to be told how to live their lives even if the benefit was clear for all to see. That was obvious from the violent reaction he got when he asked the question.

However, he thought, if he couldn't control the amount of waste, perhaps he could make it someone else's problem by spreading it around. There was after all the beautiful river Euphrates nearby carrying all manner of things away with its enormous power.

So, quietly at night, he began tipping refuse directly into the river. He knew deep down this was not really a good idea because it would end up in someone else's backyard further downstream. But if it would solve his immediate problem! It was the best idea he could come up with and the pressure was so great he experimented, first tipping a little in every day and then gradually a bit more, until soon he was dumping everything he collected.

As if by magic the outcry by the good citizens of Ur quickly abated. The pressure on him went away and he found himself richly praised by the Council and rewarded for his good work.

"Problem solved," people said to Ashok often enough that he had begun to half-believe it himself. But could it really be as simple as dumping the problem onto someone else? He knew in his heart it wasn't.

One day what he feared might happen, happened. Not quietly and out of sight, but dramatically and violently, with no place to hide.

The villagers downstream didn't at all appreciate the rubbish washing up at their doors. Once they realised it was coming from Ur they got extremely angry, and decided to send a delegation to the Elders demanding it stop.

To show they meant business they sent soldiers as well in the night and took hostages from among the leading citizens of Ur to show just how angry they were.

"Stop putting your rubbish in the river," was their message to the people of Ur, "or you'll regret it. Some of the hostages are old and poorly but we won't release them while you continue to abuse our rights like this. They are well hidden and will remain hostages with us for as long as it takes, however much bad publicity we get and whether you like it or not."

The problem of refuse had re-appeared with a new twist.

Ashok's problem was now worse than ever. What had seemed like a good idea had turned out to be a very bad idea indeed.

The citizens of Ur began complaining vehemently about the rubbish once again clogging up their streets, and now they complained as well about the return of their friends and relatives who had been taken hostage.

They made it clear to the Elders just how unhappy they were. The Elders had a scapegoat and in turn vented their anger on to Ashok who moved from hero to villain of the piece overnight.

Poor Ashok. He had tried to solve the problem by off-loading it but that hadn't worked. He had to think again, and think very quickly.

But, try as he might, he could not find a solution. He felt helpless and at the same time responsible. "I have been given this trust and it turns out I am not up to it. I am a hopeless dustman," he concluded.

He retired to the little room he had found and asked himself again and again, "What must I do?" His head ached as he tossed and turned in bed that night. If this was part of his calling to Ur, he had obviously failed miserably. For comfort he fingered his amulet and at last drifted off in an uneasy sleep.

In his sleep that night he had a dream and in the dream a man leant over him.

"I write music," said the man. "My name is Arnold Schoenberg and I've come to you from the future. I'm going to help you."

"How?" asked Ashok in his dream. "Nobody can help me."

"Look at this pencil," said Schoenberg. "I use it to write music."

He held out the pencil. It had a sharp lead for writing at one end and a rubber at the other.

Arnold Schoenberg pointed to the end with the rubber. "This end is more important than the other end," he said. "Think about it."

"Well," Ashok remembered thinking. "That wasn't much use. What am I supposed to do with that?"

"this end is more impartant than the other"

When he woke up the dream was still very vivid. Ashok was still pondering what it could possibly mean when he went out on his rounds the next day. Dejectedly, he picked up what he could from the piles left outside each house.

As usual he passed by the houses marked with a red cross as the sign of the plague without a second thought.

But today he felt a little light go on inside his head, a glimpse of something that felt might be important. For the first time he realised consciously that he never collected rubbish from them because they put nothing out. And they put nothing out because they took nothing in. Fearful of contracting the plague nobody would go near them. They consumed nothing brought from the outside, only what they could grow for themselves.

With mounting excitement Ashok exclaimed out loud, "Aha! Nothing in, nothing out!" The problem of excess rubbish was rubbed out in their case before it began.

He could suddenly see he had been looking at the wrong end of the problem. Just like the pencil it was the rubber end which suggested a solution.

"Thank you Mr. Schoenberg" he thought. "Now, how can I put that simple idea to work?"

AHA

He thought about that all day and that night, as an experiment, avoiding those who put out only small amounts of rubbish each day, Ashok began marking the doors of those who put out huge amounts with a red cross. "Let's see what happens," he said to himself. "I can always rub out the crosses if it doesn't work."

But it did work, it really did. Sure enough the worst offenders soon stopped putting out rubbish just like the houses bearing the genuine sign of the plague.

They could no longer indulge in conspicuous consumption because no one wanted to deal with them anymore. They too, in the same way as the real bearers of the plague, were ostracised. The pile of rubbish he had to dispose of was greatly reduced.

For the first time Ashok felt he could relax a little if only temporarily.

"Thank you once again, Arnold," he thought to himself. "We now know we could solve the problem if the Elders regulated it in everyone's best interest. Or, better still, if everyone chose willingly to wear a little red cross of restraint."

"But how can we do that?" he wondered. "I can't get Arnold to visit everybody in their dreams. Or can I?"

Legend goes on to say that after thirty-two days the citizens of Ur had a sudden, communal realisation. The streets they had previously avoided were now pleasant to walk in and they took to strolling and sitting outside in the cool of the evening, chatting together as they had done before. The unsightly piles of rubbish had disappeared and the all-pervasive smell of rotting garbage no longer hung over the city.

People opened their windows and called to each other in passing. They became once more a community.

Nudged by Schoenberg's idea in the dream, Ashok had discovered something so obvious he couldn't quite believe it hadn't been thought of before. The less you consume, the less you waste. And the people of Ur clearly liked the results.

Ashok once again became a figure of acclaim. He felt excited at discovering the problem could be turned upside down and resolved. But, by now, he was wary of how easily things could go wrong.

What would happen when he erased the signs of the cross he had marked on the doors of the big consumers? He knew they hadn't the authority of the Elders in Council behind them. Would they have to be made mandatory? Or would the citizens behave differently and constrain their consumption voluntarily?

He fretted that once released from their enforced abstinence they had would revert to their previous sense of entitlement to consume as much 'stuff' as before, and fill once more the streets with piles of rubbish and the air with toxic pollution.

But an extraordinary thing had happened.

When Ashok rubbed out the crosses he had put on their doors, and those who had been temporarily deprived of 'stuff' could indulge themselves again as much as they wished, their tastes and behaviours had been transformed. When they emerged from isolation, it was as if to a new land from exile.

A new idea had taken shape in the heads of most of the citizens. They liked the way the city had become cleaner and more pleasant as a place to live. The mothers were particularly pleased for their children and formed an association to protect the environment. People began to do things together outside rather than retreat inside the walls of their houses.

Even more extraordinarily they took to arguing in the forum for the advantages of consuming less and wasting less. They praised nature's rivers and forests and argued for more dustmen like Ashok.

They became the new thought leaders and Ur prospered.

NATURE

Ashok overnight became a popular hero. By general acclaim he was invited by the newly formed Council of Elders to lead the Association for Wellbeing. The Gods of the rivers and rocks and streams were celebrated in great festivals of thanksgiving.

The Environment healed and the people prospered.

Ashok in his private moments was secretly pleased, but he was worried that the public acclaim overstated his role. He had made some unusual connections and showed a way forward. But it had been a communal epiphany which had done the trick, energising the will to make the transformation of values and new ways of being together in society possible.

The people had granted Ashok the gift of trust. But he continued to insist he was just a dustman, who was in the right place at the right time and had the good fortune to solve a big problem.

But it was to turn out that he would be much more than that.

He was offered and politely declined every kind of public office in Ur, retiring soon after to a hut in the desert nearby, where he lived the simple life of a hermit.

However, fame followed Ashok and his reputation continued to grow.

From lands far away people came to visit him to seek his wisdom. They brought with them many problems to which they had no answer. Problems of drought and fire, violence and social injustice.

They found some comfort in sitting long into the night talking amongst themselves about their problems. Particularly they talked about what kind of society they wanted their children to grow up in.

They hungered for Ashok's wisdom.

Ashok spoke very little at these assemblies, occasionally jotting things down on the wall of his hut which began to resemble maps which suggested ways to move forward.

When he did speak the travellers listened intently and marvelled amongst themselves at how he made connections between things which seemed initially so far apart. Things that he plucked from the discussion and chalked onto the pathways as they were spoken made sense when they were joined up.

When asked about the ways of creativity and conciliation he spoke seemingly in riddles. He praised the gaps between things as the source of their energy. That was where the possibility of accommodation took shape. Teaching was not talking but listening. Differences had to be celebrated before they could be bridged. Finding common purpose came from going down before a different idea rather than confronting it.

But out of these apparent riddles people found their imagination had changed. Things previously apart and in opposition came together. Winning became less important than understanding. They found the will and the way grew together.

He became known far and wide as Ashok the Wise, Guardian of the Circles of Wisdom, Harmoniser of Conflicts, Teacher of the Way.

When the travellers left for home they took with them the idea that by talking together and listening more carefully to each other they could solve problems better before they happened rather than afterwards.

They had stumbled at first over how to listen and who should be given voice. Gradually they came to realise that truth was a many sided creature which sometimes resided in the mouths of enemies and people who were easily ignored.

And that a stronger sense of purpose could emerge from waiting and listening than rushing to action.

When they returned home they found their communities flourished when they practised what they had learned in the desert. Others wanted to experience what those who returned spoke so joyously about. In this way ideas about common purpose migrated across great distances and changed the way people thought and behaved.

More and more people made the journey to Ashok. He in his turn lit a candle of respect for everyone who passed through, and his place in the desert became a cradle of light and a place of pilgrimage.

Some problems, however, were so huge and far beyond the experience of those who came together in the desert, they seemed insoluble.

How could love flourish when there was injustice and inequality? How could a man find peace when his context was competition between individuals for wealth and countries were in constant preparation for war? How to prevent the extreme events of floods and wildfires and the oceans dying from global warming?

Ashok knew he had no ready-made answers. His experience was limited to the desert, to Ur and the narrow problem of waste. He began to wonder whether his calling had not just been to Ur as at first he had thought. Was that why his mother, Gaia, had looked with such penetration into the very soul of his eyes when he told her of the voice which summoned him in the night.

The only way to find out was to leave his place in the desert and see for himself what was going on elsewhere in the world. What was driving these problems? Why was beauty so disfigured by evil? Why was hope so riddled with despair? Had the threads of a better common good ever been discovered in different ages and far way places? Could they be rewoven into a cloth of gold?

To find out he would have to travel as a listener not as a teacher, and find out for himself.

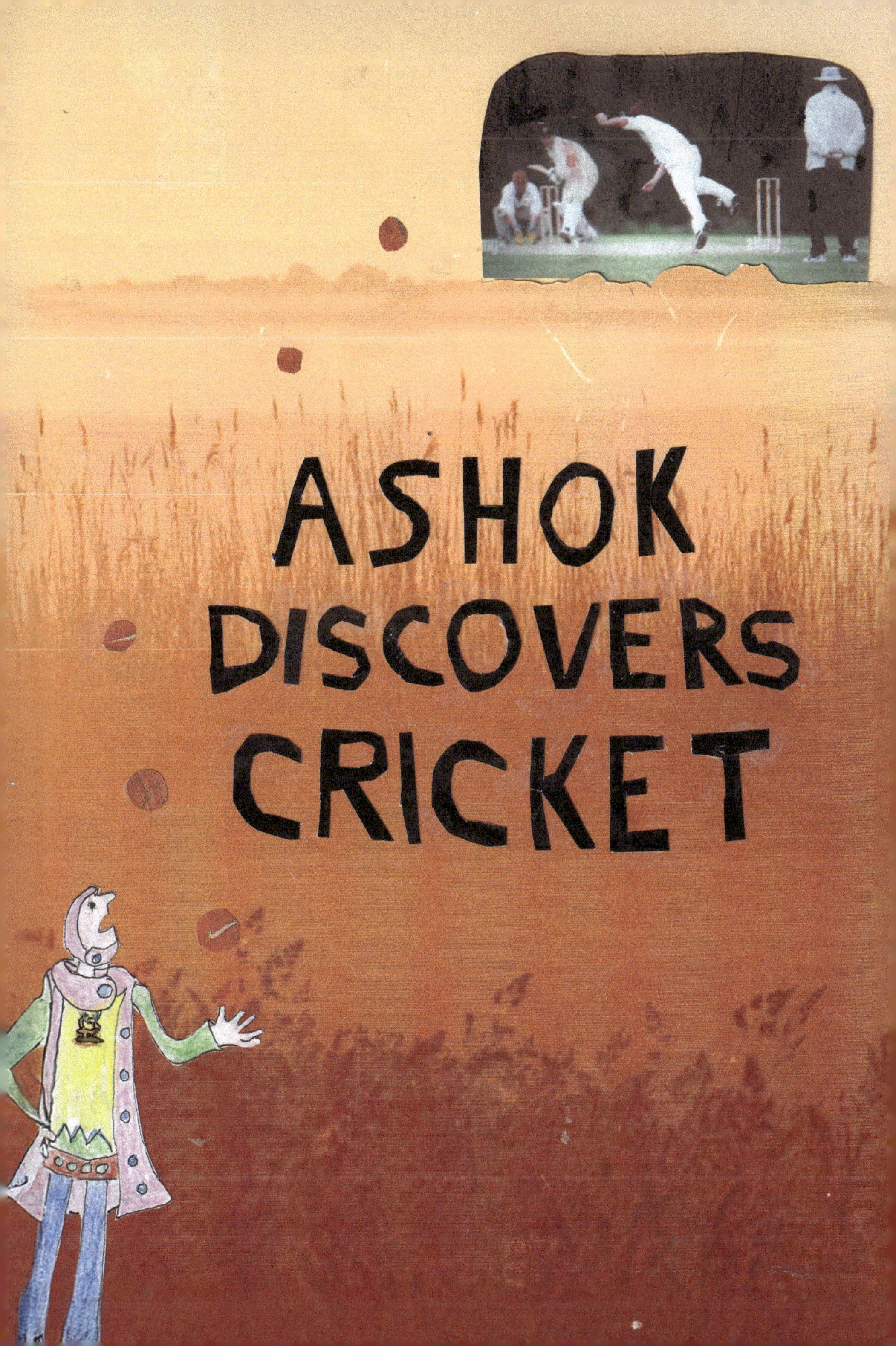
ASHOK
DISCOVERS
CRICKET

Ashok, who had come out of the desert to become the First Dustman of the modern world at a place called Ur, had become famous throughout the ancient world for solving really tricky problems. In the case of Ur it was too much waste being dumped on the streets as more and more people consumed more and more stuff and the city grew bigger and richer.

He had then withdrawn from public life and retired to a small hut in the desert, but his reputation for wisdom drew people from every corner of the world to share their problems with him in the belief he could help.

He became known far and wide as Ashok the Wise, Guardian of the Circles of Wisdom.

But he was a very humble man who had been called to Ur by a voice from the sky. The problems of conflict, migration and climate catastrophe which were now brought to him were beyond his experience. He knew that, to be helpful, he needed to go to where the problems were most acute and see for himself what was happening.

Luckily, when he had a problem he needed advice on, he could speak with his all-seeing mother Gaia who had taught him that by concentrating intensely and emptying his mind, he could heighten his consciousness by meditation.

In this state, the laws of physics fell away and he could travel to anywhere in the world and drop down through wormholes in time to experience how life had been in ancient cultures and times long gone.

Fingering the small amulet she had given him when he had first been called to set out on his journey, and which he had worn constantly at his neck ever since, he was able to travel effortlessly anywhere into time and space.

Today he had found himself in the dry, high deserts above the beautiful city of Muscat in Oman. He had seen and marvelled at its tranquil bay and the beauty of its buildings by day, but now he was tossing and turning in the middle of a terrible dream.

In the dream he could see in his mind's eye a gigantic wall of water building up above the city, poised to crash down from the mountaintops onto the unsuspecting inhabitants below. He watched it surge forward carrying everything before it.

He could feel its irresistible force as it burst through the souks in the narrow streets lining the bay, emptying their treasures of gold and frankincense and myrrh into the sea. The dhows plying their wares along the waterfront were tossed in the air like so many splinters. Even the great mosques and palaces shuddered under the impact of the torrent.

People witnessing the devastation from high places of refuge saw lifelong friends and relatives carried irresistibly away in the turbulence. They wailed and lamented, fearing it was a sign that the Gods had been angered, and worried what would come next.

Ashok could feel their fear and sorrow in his dream as if it was his own.

In the morning when Ashok awoke, he was fearful of the destruction he would see. But, to his amazement, everything was as it had always been.

The sunlight from a clear blue sky reflected off the Sultan's palace and the minarets of the mosques, just like it had every day. In the bay colourful dhows criss-crossed the still waters plying their goods as they had the day before and the day before that. Merchants in the souks fingered the beads of their misbaha and gave thanks to Allah that all was well.

Ashok was relieved to see that the flood had only happened in his head, which now ached with the ferocity of its memory. But was this a premonition of things to come? Had he fallen through a gap in time and seen events that were still in the future, as Schoenberg had done in his dream at Ur?

He spent the day unable to rid himself of the terrifying prospect that he had been given a glimpse of things yet to come that were beyond control.

That night his fears were heightened when travellers from far and wide came to sit by his fire. There was a greater sense of urgency and desperation amongst them than ever before. They told tales of floods and fire in the lands they had travelled through, the like of which had never been experienced.

They had seen panic and fighting in the streets, and frightened voices questioned why this was happening and what could be done. Having seen the horrors with their own eyes, people were angry. They demanded action before it was too late.

There was fear that nature had turned against them. Fingers were pointed at newcomers and persecuted minorities and angry scuffles broke out amongst the restless crowd. The militias were mobilised to keep order in the streets and imposed curfews in an attempt to prevent full-scale riots. Civil authorities progressively gave way to the military.

Towards dawn, as the hubbub gradually died down, a man detached himself from the crowd and sought out Ashok.

He explained he was the captain of the large warship at anchor in the harbour of the bay. He had heard tell of a wise man in the desert, who was said to possess the gift of helping people to solve unusual problems, problems that were too complicated for individuals or even governments to manage on their own.

He had searched far and wide and been led to Ashok. His own country, although far away and more developed than Oman, was experiencing exactly the same terrible floods and fires the travellers had recounted that night.

Catastrophic events were becoming more and more frequent, and people in all walks of life were becoming more and more frightened. They were apprehensive that the balance of the natural world had been disturbed, and their governments weren't doing everything they could to appease nature.

He said, for many years, concerned and thoughtful people from many countries had met together to find ways to stop the remorseless rise in temperatures occurring everywhere across the world.

Nowhere was felt any longer to be safe. The land and the oceans and the ice caps were heating up, creating catastrophic heat domes and rising sea levels. People were dying of famine, drought and warfare. And people were afraid.

Numerous small initiatives had been taken which were widely publicised. But far from solving the problem they fell further and further behind. People were becoming increasingly desperate that, despite the endless talking and the many promises, nothing of substance was being done which really got to grips with the problem. It was as if humankind was sleepwalking to extinction.

The resistance to any radical plans to reduce consumption, energy use, and growth was so strong the belief had become commonplace that everything would work out for the best in the end. Things may look bad, but science and the ingenuity of man have always prevailed. Philosophies had even developed which held that staying the course we were on was not only the only practical option, but the right moral course to pursue.

As it became more and more obvious that political leaders were not going to step up, attention turned to the risk of irreversible hazard, from which there would be no way back. People wanted more positive action and were becoming desperate for guidance about what to do.

Ashok felt an instinctive bond with this man from the other side of the world, so he willingly accepted when he was invited to travel with him back to England. "Come and look, with the new eyes of a man from the desert, at what is happening in the big, rich countries which have been at the forefront of economic development for many generations. Help us to find solutions which work."

This was an invitation that Ashok felt he could not refuse.

And so it was, the very next day he found himself journeying down one side of Africa, rounding the Cape and sailing back up the other as a guest on one of His Britannic Majesty's warships.

This was all new territory to Ashok and he was enchanted by everything he saw. He was particularly excited when they put into various ports along the way, following a time-honoured tradition of playing cricket against a local team.

Little did he know how important this was to be.

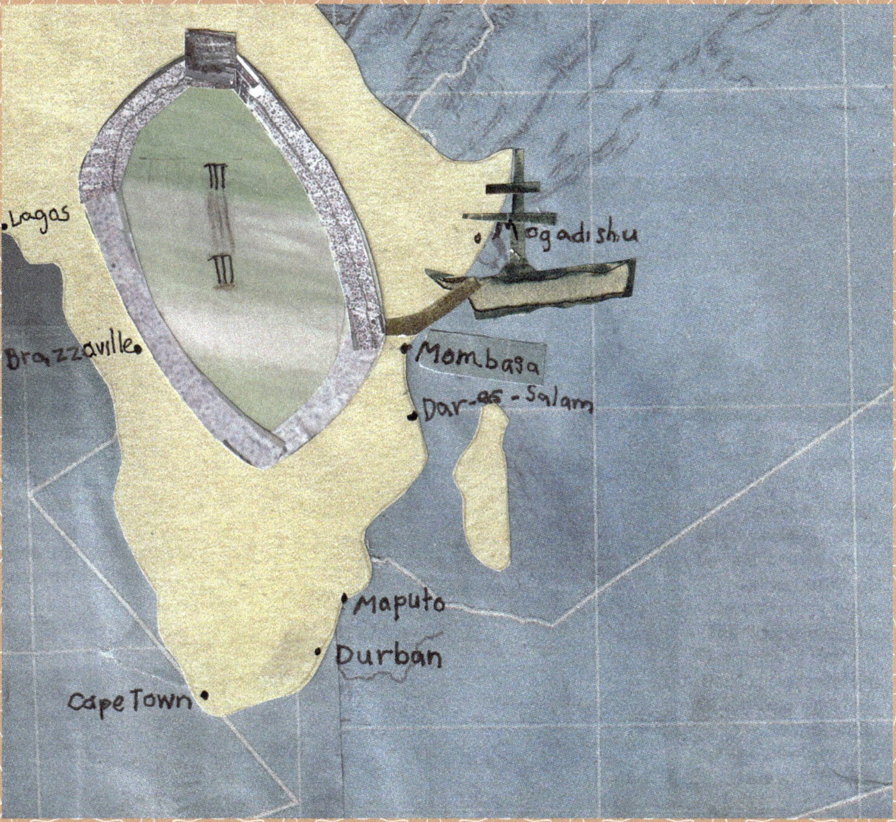
Lagas
Mogadishu
Brazzaville
Mombasa
Dar-es-Salam
Maputo
Durban
Cape Town

Now cricket was new territory for Ashok and, try as he might, he couldn't figure out why the ship's team invariably won each match they played by such huge margins. Over dinner in the mess he asked why that was. Why, in what should have been an even contest, was there never any doubt about the outcome?

This was met with much banter and laughter and when it died down the Captain explained to Ashok it was really quite simple. Cricket is a gentleman's game and both sides play by the rules and jealously respect them. The difference is we have all the star performers on our side. Lumpy here takes all the wickets, Bullen scores tons of runs. The rest of the team take the catches and chase down everything in the field.

"In short," he said, "we bring more firepower to the contest than our opponents within the rules of the game. And so we win."

"Aha," thought Ashok, "that's exactly what happened when the hostages were taken at Ur. The strongest force won, but it didn't solve the problem."

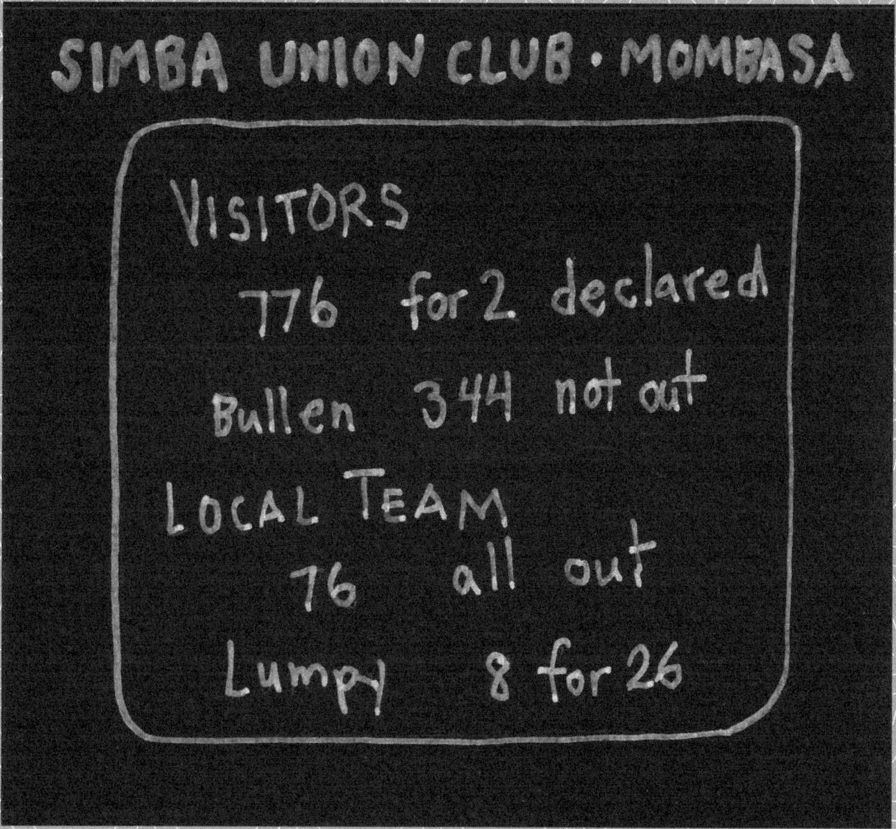
SIMBA UNION CLUB • MOMBASA
VISITORS
776 for 2 declared
Bullen 344 not out
LOCAL TEAM
76 all out
Lumpy 8 for 26

Once it had been pointed out to him, Ashok could see for himself the mismatch of strengths between the teams as they put into romantic-sounding places like Mogadishu and Brazzaville and Freetown. The local team played whoever was available on the day. The ship's team were well rehearsed and had the firepower of Bullen and Lumpy, when it really mattered.

Later, from the window of the train which took him up from Portsmouth to London, he sat idly watching the endless stream of cars criss-crossing the countryside, and the smoke coming from the chimneys of large factories next to the railway line. He thought to himself the vehicles were a bit like the dhows at Muscat, but so much noisier, tail to tail, leaving a corridor of fumes rather than foam in their wake. And whilst the towering chimneys resembled the minarets of the mosques, they gave off black smoke rather than prayers to God.

As the train approached London the traffic got denser and slower, as if running on sand which got softer and deeper. The houses seemed to join up to make one continuous building, with tiny patches of green and brown at the front and the rear as far as the eye could see, each sprouting identical aerials on the front and chimneys on top.

Once installed in a magnificent hotel on the Strand, he wandered this way and that amidst the hordes of tourists jostling in the streets and spilling out of restaurants and bars. They reminded him of the traders with their caravans of goods for sale arriving in the already bustling city of Ur, except these tourists were obviously here to have a good time.

That night he found he couldn't sleep in his room because of the powerful neon lights opposite advertising luxury pet foods, cheap flights and a bewildering array of trainers that, in an idle moment, he worked out would cost the equivalent of a year's salary in Muscat.

He felt oppressed by the constant noise and activity of the big town and found himself yearning for the stillness and silence of the desert. To fill time before dawn, he took to sitting in the foyer watching the endless traffic go by and marvelling at the cardboard camps which formed, as if by magic, in the doorways opposite when it rained, like Bedouin seeking shelter from a storm.

HOTEL

At breakfast he was astonished by the array of foods, each nestling under its silver dome like a little private palace, and, even more, by the sheer weight of the newspaper which was brought to his table.

"So much paper," he thought, "just to describe what was going on in the world." He calculated idly that a mountain goat could survive for five days on the trees cut down to make just this one newspaper.

The front page was dominated by pictures of floods covering miles of low-lying land, while groups of bewildered people stood around in tears contemplating the destruction of their new homes. On the same page was the announcement of new drilling activity for oil and gas in the North Sea.

He smiled to himself at the irony of the juxtaposition of the two events as if they were completely unconnected. "How many new floods and fires he wondered would be caused by those two small announcements sitting so innocently side by side?" he wondered.

FLOODS
TIMES

Back on the ship, over the reunion dinner, the Captain called for silence and asked Ashok if he would like to share what, if anything, he had learned after a week watching a leading economy going about its daily business of work, rest and play.

People looked at each other in embarrassment. They had got to like Ashok, but what could this little man from the desert possibly contribute where the best minds in the developed world had so consistently failed?

He started by admitting he had felt quite lost when he saw how the worlds of economics and politics and behaviours were so big and entangled, like the desert sand and its winds.

Then for some reason he found himself thinking again about the wonderful game of cricket, and something flashed in his mind. He realised he owed his colleagues a debt of gratitude bigger than they could possibly imagine. They had given him a way of thinking about the problem of climate change and its solution which he wanted to share with them now.

Intrigued, everyone at the table began to pay more attention. Cricket as the solution to climate change? He was either going to pull a rabbit from the hat, or fall flat on his face. Whichever, they were now listening carefully.

"Think of it for a moment," he said, "as if it was a game of cricket. For thousands of years we burned fuels from the earth to keep ourselves warm, cook our food, and keep the wolf from the door.

"Then, with the ending of the ice age we could leave our caves to hunt and harvest. Life became so much easier. In celebration we multiplied but we still weren't many, and for thousands more years we were very junior partners with nature, with the forests which still kept growing and the rivers which ran clean.

"We were small and nature was big."

"Then about 300 years ago – the blink of an eye in the long cycle of time – the balance suddenly and dramatically changed. We learned how to generate steam power from burning coal and how to make and move and consume things quicker and more cheaply than ever before.

"And now we began to multiply for serious. We grew in numbers beyond imagining, and every generation, every year, every day we wanted more and more 'stuff'. Stuff to consume, stuff for our factories, stuff for play.

"Our appetite grew by eating.

"Overnight a new way of thinking and organising came into existence driven by this appetite for more of everything, more consumption, more innovation, more advantage, more profit and more accumulated wealth.

"It was a revolution in the way we related to each other and to nature, which we no longer treated as a partner but a resource to be exploited for our benefit."

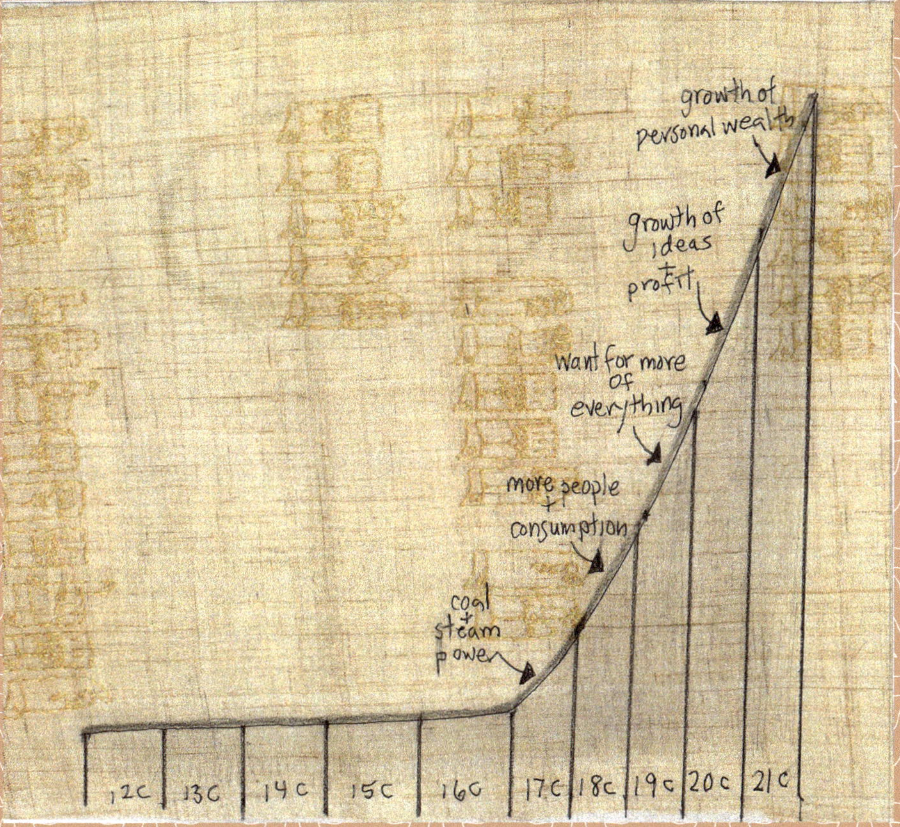
growth of personal wealth
growth of ideas + profit
want for more of everything
more people + consumption
coal + steam power
12c
13c
14c
15c
16c
17c
18c
19c
20c
21c

“Something fundamental had happened. The game changed from ‘being’ to ‘having’. Growth of consumption and competition between individuals and states for the accumulation of wealth became the dominant trend of our time.

“Governments and businesses the world over combined to re-write the rules of society in pursuit of growth. All kinds of institutions and incentives came into being to help win at the new game. Banks, credit companies, advertisers, newspapers, traders, retailers, politicians, companies, lobby groups and think tanks, all became big players in the economic engine designed for growth.

“And they all had a vested interest to keep it motoring along.

“For the very first time humans had the winning team against nature and, in their billions, they hacked away bit by bit at the atmosphere and the natural systems which govern life on Earth and keep things in balance. Even now, as the world heats up and we can see the consequences of it getting more and more angry, the dominant players defend growth and accumulation as the ‘common good’ which results from everyone doing what is best for themselves.”

BEING
HAVING

Everyone present was now listening intensely to the little man from the desert.

"It's like the Captain said," he went on, "it's really quite simple. At cricket you win because you have the strongest team. When it comes to playing the big 'green match in the sky' against nature, all the strongest players are on the growth team: companies, banks, technology, consumers and governments – the equivalents of Lumpy and Bullen – all play for the same side. And they can make the rules what they want them to be.

"So a few individuals can own half the world's wealth. The surrounding pack want their share. Even the relative losers demand their day in the sun and get behind the wheel whenever they can, because they can. Holidays are an entitlement. Taxis keep their engines running on the rank. Ghost flights without passengers take off to preserve landing slots.

"Pollution and environmental destruction grow as fast as consumption.

"The reality is we do not organise ourselves to heal the problem of global warming, we organise ourselves to bake the problem into our daily lives in the conviction this is the best of all possible worlds."

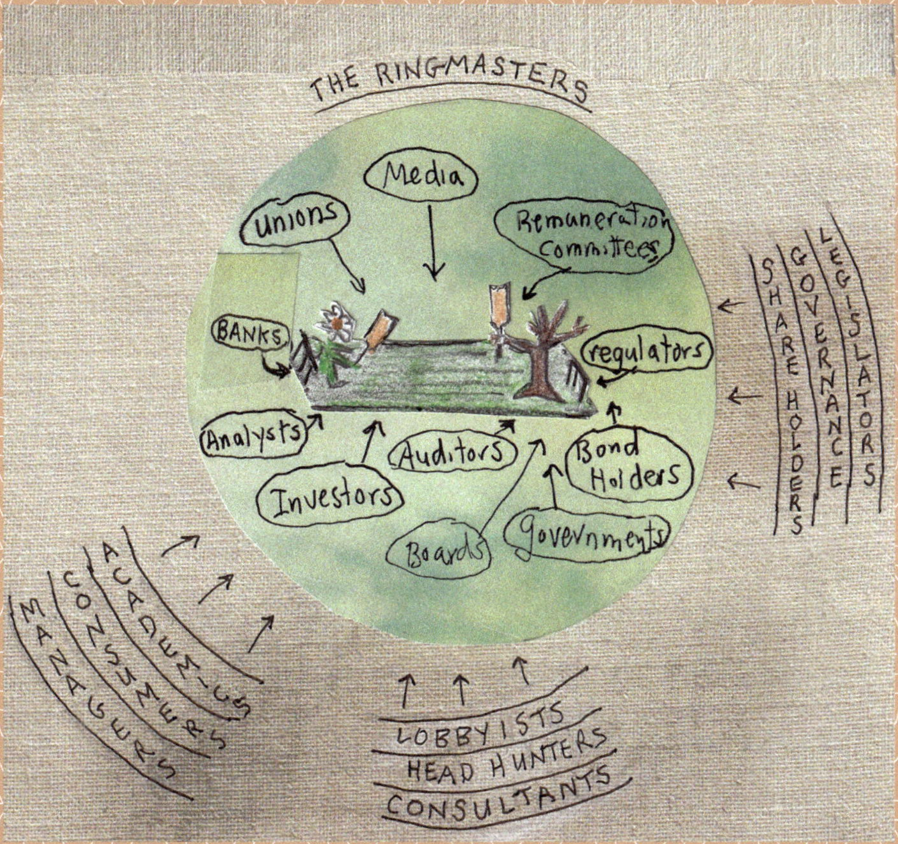
THE RINGMASTERS
Media
Unions
Remuneration committees
BANKS
regulators
Analysts
Auditors
Bond Holders
Investors
Boards
Governments
SHARE HOLDERS
GOVERNANCE
LEGISLATORS
ACADEMICS
CONSUMERS
MANAGERS
LOBBYISTS
HEAD HUNTERS
CONSULTANTS

"Civilisations play a version of a continuous game of cricket against each other over time. They compete to get the largest possible share of resources. The more power they have the more resources they can command. The winners build monuments to celebrate their success, some brash and transient, some enduring and spiritual.

"The monuments the growth team has built are killing power, disposable trash, ever bigger newspapers, twenty-four-hour neon lights, constant TV advertising, incessant promotion of consumption, debt so deep we lose sight of it, unbalanced reward systems, destination travel as one item on a tick list of entitlements.

"The monuments to growth become larger daily as governments relinquish the high ground of purpose and values, and compete to curry favour with electors who increasingly view the world through the lens of consumers.

"When the boundary proves too long for the home team to hit the sixes they are paid to hit, the rope gets pulled in a few yards to keep the momentum going. And in the field the lobby members put in the legwork to ensure the rules remain forever favourable."

Now the silence was so intense you could hear a pin drop as Ashok drew to a conclusion.

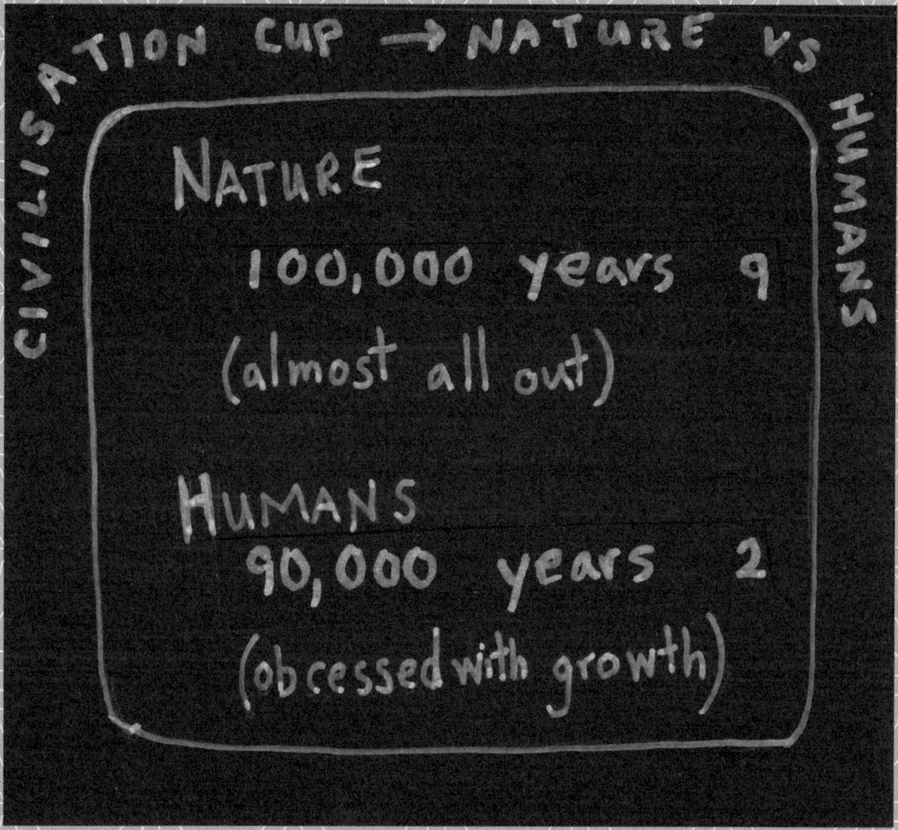
CIVILISATION CUP → NATURE VS HUMANS
NATURE
100,000 years 9
(almost all out)
HUMANS
90,000 years 2
(obcessed with growth)

"What people in the street and on the buses have told me in my short stay is that we pay lip service to honourable objectives like stability, security and sustainability, but they are not the monuments we actually revere.

"Our monuments suggest we revere the analytic ability to break things down into their smallest parts, the ability to find financial value in those parts and to trade them as often as we can, and the ability to accelerate, market and globalise those fractions.

"We have simultaneously lost the ability to see the whole of a thing in its context, the ability to see the intrinsic value in things beyond their monetary value, and the patience to build relationships and trust instead of trade.

"As a result, we are playing a game of chicken with nature. And it's one we can only lose. In the name of individualism and freedom we endorse the very icons which are destroying us. To have any hope of combating climate catastrophe we need to fashion a completely new idea of society based on the interconnectivity of everything in nature, the slowness of community, the importance of consciousness, and the power of relationships and co-operation.

"What they say is they are ready for a new story with new values, new institutions and new leaders, but have no idea how to stop the bus of growth they are currently on."

ICON
REVIEW
Travel posters
24 hour neon lights
unbalanced reward systems
Constant TV advertising
government promises
FAT Newspapers

There was a deathly silence as Ashok resumed his seat, followed immediately by a riot of noise as the import of what they had heard went home. The Captain turned towards Ashok, but he was no longer there.

Teleported back through the ether to his place in the desert, he was already watching with horror as the nightmare he had dreamed about unfolded in real time before his eyes. The heavens had opened and a giant wall of water was rushing past him to engulf the city below.

With his head in his hands, Ashok realised that understanding the problem was not enough. He had to discover how to plant a new story about love, co-operation, community, place and patience in enough minds before it was too late.

How, for heaven's sake, could he do that?

ASHOK
WONDERS
WHETHER
HUMPTY
DUMPTY
CAN BE
PUT
TOGETHER
AGAIN

The citizens of Ur had come to think of Ashok as the wisest man in the world. His reputation had spread far and wide and travellers now came to him from all points of the compass for guidance when they had difficult problems. They took comfort from the way he listened to their concerns and found ways of connecting small threads of hope and solutions together, which surfaced as they talked.

He, on the other hand, thought of himself as a man from the desert, someone who cared for the winds and the sandstorms and the stars that studded the sky at night. They were his friends and partners who helped keep nature and men in balance.

Recently, however, travellers had started telling stories about floods and fires and melting glaciers never previously experienced. He could feel in his soul that that fragile balance was being dangerously disrupted.

Those who came to him were clearly frightened and increasingly desperate to find answers to the climate tragedies which were happening all around them before, like Humpty Dumpty, the world fell off its wall and couldn't ever be put back together again.

Right now Ashok was watching a mountain of water rising up out of the desert in Oman, crashing down on the timeless beauty of the city of Muscat and its unsuspecting souks fronting onto the bay. He was deeply worried that the wall of water he could see raised high above the city was just one of the many floods and fires occurring around the world.

In his dream, he asked himself, "Why has the age-old balance with nature been lost? Was it ordinary people going about their daily lives who were unintentionally turning nature against them and making the climate angry?"

"Or was everything that happened in this world already written in the stars and inevitable?"

If it was pre-ordained, he would have to hunker down with the rest of the world and just make the best of what was coming.

If it wasn't, he was reluctant to accept that mankind would willingly sign up for its own annihilation. Surely for heaven's sake it could be prevented!

He consulted the ancient religions which all gave him the same answer. "Deus vult," declared the Christian world. "It is God's wish."

"In sha Allah" echoed the Muslim world. "Praise be to the one God."

"If you want to make God smile, tell him what you think you will be doing in a week's time," said the Jew on his way to the synagogue. "Only God knows."

But those answers still left Ashok unsettled. Was it really God who decreed the climate get hotter and the planet more and more angry? Was the climate mess like the broken egg that could never be unbroken? Or the mighty oak tree that could never get back into its acorn? Or Humpty Dumpty who fell off the wall and couldn't be put back together again?

Ashok knew that many beautiful and mysterious and frightening things took place in nature. Sunsets that lift hearts and earthquakes that fill morgues. Beauty and horror combined and beyond our control.

But was climate catastrophe like that? Or was it us humans who had got above themselves and imposed our ascendancy over the balance of nature? What if it was our own doing?

But, if it was, in a way how wonderful. There might still be hope if we could find the 'will' and the 'way' to defeat the storms and the heat domes that spelt death and destruction to the poor and the rich alike. After all man was gifted uniquely with the capacity for moral judgment and the genius for invention and co-operation.

Wise as he was, Ashok didn't know the answer, and when he was out of his depth, he did what children have always done. He flew as quickly as he could through space and time to be with his mother and seek her advice.

Emptying his mind, and fingering his amulet, he was transported in an instant through the ether and found himself sitting on a mat of dried grasses at Gaia's side. The evening was calm as they sat looking out onto the foothills surrounding Ur.

It was harvest time and all day long they had sat together in perfect harmony tossing sheaves of wild oats high in the air, letting the gentle winds separate the gossamer-thin pods from the seeds.

The pods fluttered above their heads, like butterflies released from school at the end of the day.

They worked comfortably in silence together but sensing his turmoil Gaia asked him gently what was troubling him so deeply. Feeling secure in her love he asked his mother, whose wisdom he trusted above all others, whether the floods and fires now happening all over the world were subject to the same iron laws of necessity as the pods that were floating away?

"Has climate disruption," he asked, "already become irreversible like the pods separating from the seeds?"

He knew that the ancients spoke of a great flood and of mountains erupting into fireballs that destroyed everything in their immediate path. But was climate change like that? Now it seemed to be the whole earth that was baking simultaneously. Scientists spoke of a limit to just how much heat the oceans and forests could store as one bad event followed another.

He knew from his short visit to London that something profoundly important had happened when ordinary people, not God, had unleashed a process they called the Industrial Revolution. Humans had been around for millions of years by then but it wasn't until about 300 years ago, the blink of an eye in cosmic terms, that they found ways to use the energy from fuels stored deep under the surface of the earth to begin manufacturing things on a big scale.

As they consumed what they made, their appetite to consume more kept on growing. They became consumed by their own passion for growth. That meant more extracting and more burning, driving temperatures in the oceans, the air and the land to rise and rise again. And so it had gone on.

The overwhelming question now for Ashok was whether the cause of the problem – an ever-increasing appetite for more 'stuff' – had become irreversible? Was the future out of control like the gossamer pods floating away on the evening breeze? Or could humans find the wit and the will to craft a more co-operative future?

They sat together in silence as the shadows lengthened and Ashok's question hung in the air for a long time. When finally she spoke she answered with her own question: "What has the past taught us?

"What would the great King Ashurbanipal, creator of the greatest empire in the world seven hundred years before Christ, say if you asked him? He had at his fingertips all the wisdom of the world stored away in his great library at Nineveh, including the *Epic of Gilgamesh*".

"Or, what about those who, after centuries of warfare throughout Japan, passed through the gateways of enlightenment in the forests of Nara and discovered the way of flourishing and peace? Or, in the modern world of financial transactions, the master gunner driven by the lure of reward to suck life with his harpoon from the legendary singing whale, as it arced gracefully above the surface of the ocean?"

Ashok knew from old she would not answer her own questions. He could only take the clues, and difficult as they were, try to solve the riddle for himself. Holding her tight he gave thanks for her wisdom and set off immediately for Nineveh, far distant in time and space, to consult with the great King Ashurbanipal of Assyria.

When he arrived he was overwhelmed by the beauty of the carved frieze above the huge library doors. It depicted a magnificent lion writhing in its death throes, vomiting blood, fatally run through on the royal spear during a hunt.

He marvelled at the strength and courage of such a great king who could confront danger like that, and respectfully asked him the secret of such will. Ashurbanipal replied proudly that it was his sacred duty to keep his subjects safe from predators, regardless of the risk to himself. Killing lions was his duty.

"Aha," thought Ashok, "he killed lions as a sacred duty." But he also knew that, forever after and faithful to the example set by the King, his courtiers in turn copied what he did and proudly killed more lions. Soon it became a badge of honour in Assyria to kill lions. The habit became a value embedded in the culture and expressed in the monuments of the empire, enduring for thousands of years, until the very last lion and her four cubs were shot in the deserts of Iran, in the year 1963.

"So," he mused, "that's how something that starts small and honourably, loses its purpose in the complexity of human organisation, and becomes an existential threat. Extinction for the lions came not by asteroid, or the will of God, but by mankind, who despite his wisdom, couldn't stop doing what he enjoyed, and felt it his duty to do, until it was too late."

The realisation that mankind could operate for so long on auto-pilot, unaware of the consequences, shook Ashok deeply. It was clear Assyria was not going to give him the answer to his problem. If anything, it had demonstrated how blindly irreversible the ways of man can be, regardless of the will of God.

So he set off once again, this time to the outer limits of the Indian Ocean to see whether the whalers, those fabled men who took such enormous personal risks to hunt the giant sperm whale, could provide him with an answer. Where did they get their will to challenge the might of the oceans?

He spoke to them on the quayside as they returned exhausted but triumphant from months at sea. These were the master gunners, seasoned professionals who took pride in their reputation for being able to hit a whale at the very peak of its spectacular leap from the water.

Many spoke to him, cautiously at first, and then more openly of the shame they felt at the agony they inflicted on the king of the ocean, as the charge on their harpoon exploded inside the belly of the whale.

Most, however, just turned their backs and shuffled off, saying it was just a job, and the money on offer was exceptional. "You do what you get paid to do, don't you, and we get very well paid for killing whales."

Ashok pondered for a long time what they said. Hunting out of necessity for food or as a rite of passage, was something he could understand. But the whalers weren't doing that. They were following the money.

And money, he knew, was completely neutral as to morals. In fact, the case for markets was they removed human judgement as much as possible from the way resources were used. If money could be used to motivate people to do bad things as well as good things, where might that lead? What kind of civilisation might that produce? What would be its enduring monuments?

Could it be money which was driving the whole world towards a spectacular end, which nobody but the odd doomsayer consciously wanted, by damaging the harmony of man's relationships with nature? Had the pursuit of accumulating wealth for its own sake unleashed a dark force, hitherto anchored in the soul of humankind?

Was this the insight Gaia had wanted him to get from the whalers about what was making the world totter ever faster towards the edge of disaster?

If it was, he felt even more deeply troubled now after his encounters with both Ashurbanipal and the whale hunters. So it was with a heavy heart he set out to visit Nara, the last destination suggested by his mother.

As he entered the great forest, untouched by human hand since the beginnings of time, he was humbled by a sense of having entered a new and sacred world, as if standing on the threshold of a new level of consciousness and understanding.

This was where the Shinto in ancient times had finally laid aside their everlasting obsession with competition and warfare in exchange for the growth of a shared spirit.

He felt simultaneously very small and yet bigger in the peace and stillness of everything he encountered. Temples reached for the heavens with a sense of ascending tranquillity. Sacred deer shared the woodlands with humans and moved freely between the Gods and man, acting as messengers of the concern felt from the world beyond for all living things on earth.

These had been exceptional men, the ancients, who had learned anew how to come together in a strange land and co-operate, turning consciously away from the exile they had been living of desire for worldly domination and the possession of things, to a state of all-belonging and all-being.

Ashok passed through the Gateway of Enlightenment and experienced the softness of an all-embracing love settle on his shoulders and guide him forward.

His feet followed the path which led away from the way of life on earth, which ends abruptly in death, and returned through the Gateway of Death on the pathway to a greater life.

Everywhere and in everything he felt a deep and lasting reverence for the small, unobtrusive things and the threads which combined to make the common good.

He stood for a long time, his whole being transfixed, before the Temple of Flourishing Law, the most beautiful form of the unity of the whole ever to issue from human imagination, a monument to the omniscience of love and the growth of dignity and respect.

As he looked in wonder, he felt his field of vision physically broaden, no longer restricted to the one eye shared between the three witches who confronted Perseus. He experienced the strange sensation of being able to see with his heart, and felt his soul rise at the glimpse of the wisdom and purpose which lies beyond human knowledge.

Filled with hope, he found himself wafted once more back to his mountaintop in the desert, and was brought abruptly back to reality.

The towering wall of water he had seen in his dream was, even as he watched, smashing down on the unsuspecting port below.

He could see now with greater clarity how the war on nature was provoking the most savage response. He had witnessed on his travels to rich lands in the north how the progressive poisoning of the atmosphere had converted the sun from a partner with humankind in the provision of heat and light into an arch-enemy of arson and destruction.

The overwhelming question forced itself yet more powerfully into his mind. "Had Humpty Dumpty already fallen and broken?" If the momentum to climate catastrophe was not yet inevitable and might just be reversible by human action in the spirit of Shinto reconciliation, what must he do?

And, like all fundamental truths, the answer came out of the question.

Ashok saw how the lion king's sense of duty had led directly to the extinction of a species. And how the logic of financial markets, lacking the power of an inner eye, had decimated the whale population. But he also saw that neither were irreversible processes like the breaking of an egg, which could never be unbroken. This was more like cultural sleepwalking by humankind knowing all it needed to know to act differently. But still didn't.

For Ashok the Temple of Flourishing Law stood witness that the scales can drop from timeworn eyes, and the will can be summoned to find a new way. He knew the momentum was so strong it would require a powerful new story about the harmony of 'being' and 'having', and a seismic shift of consciousness, to achieve a similar transformation.

But Ashok no longer felt pinned and wriggling to the wall by the wrong question. The choice between leaving the common good to emerge from individuals acting in their own self-interest, or finding it exclusively in a shared communal vision, was a false one. He could see now they were not opposites.

'Being', was not something incidental to life. It was individual life and shared life joined together in common cause, private goods and public goods confirming each other in the fullness of aspiration.

Being was the prerequisite for 'becoming' and the pre-condition for rebirth and renewal.

MEANING
HARMONY
SERVICE
BEING
BECOMING

By contrast, when he thought about the great Temples of Reason that man had constructed – Science, Technology, Economics and Politics – he found them uncomfortably bare of meaning and moral purpose.

They were vehicles for the intellect, that part of human consciousness which excelled at the discovery of why things in nature were as they were. But vehicles which if you weren't very careful, could take you to places you had not intended to go. Perhaps that was the way to think about climate catastrophe, and warfare between countries and competition for more stuff.

Episodes of purposeless drift in the moral condition of man, driven by one-eyed logic as opposed to the logic of the heart. Outcomes of systems with 'unconscious' and 'unaccountable' power which had embedded themselves like scar tissue in the patterns of our thought, in the way we organised ourselves in society, in the values we held, the monuments we built and the behaviours and icons we celebrated.

How many people, Ashok wondered, would actually choose the pollution and sickness, the alienation and the obesity, the conflict and inequality which were consuming the world, if they had the choice?

Doubtless some who were benefitting from how things were, but no wonder it felt to so many a bit like being on a driverless bus for which you hadn't bought a ticket, and from which you couldn't find a way to get off.

KNOWLEDGE
POWER
SCIENCE
ECONOMICS

Back on his mountaintop in Oman, Ashok realised that Nara had taught him one thing above all others. The loss of harmony between mankind and nature was not inevitable. It was the endless pursuit of growth and power that had baked into human organisation the values of competition, the specific, the instant, the marginal and the ephemeral.

They were the values of an Enlightenment which had lit many candles of wonder and awe. But they were also candles which had put us on an extinction mission we hadn't signed up for, one perilously close to its final stop.

Our one-eyed view of progress as entitlement to material comfort had overwhelmed our capacity to pause, reflect and connect. Leadership had become event management in the here and now, naïve about context, fearful of vision, mindful of self and mindless about service and accountability.

Ashok held his head in his hands as he pondered what had to be done to stop the bus plunging mindlessly off the cliff.

Ages ago he had stumbled naively out of the desert and helped to create a small epiphany in the city of Ur, around the single problem of waste. To build a new world, when leaders and their institutions saw it as their duty to compete for more power and more growth, while ordinary people felt hope and purpose trickling like sand between their fingers, would need a world-sized epiphany.

Finally he knew why he had been called from the desert.

HOPE
PURPOSE
FAIRNESS

ASHOK
AND
THE
DESERT
SCROLLS

Ashok, fresh from his journey to England and his no-brainer realisation that, just as in cricket, if you fielded the strongest team in the battle of ideas, you were pretty much guaranteed to win, was now back on his mountaintop look-out in Oman.

The clash of civilisations and the mass migration of millions from poverty and conflict were fearsome challenges but they were within the human domain. The most fearsome battle of all was the one man had engineered with the external forces of nature, and he watched in horror as a giant wall of water smashed down through the tiny streets of Muscat, carrying off the treasures of the souks, and with them the lives and dreams of this beautiful city.

He realised that what he had seen previously in his nightmare had become a reality, taking place before his eyes. It was as though he had been given beforehand a look through the fabric of time, beyond the waking comfort of life going on as usual.

He recalled vividly the morning after his nightmare how the bay had been tranquil and the dhows criss-crossed the harbour almost playfully, as they had done from the beginning of time, plying their trades and sharing their news.

Now there was death and destruction and turmoil everywhere, as people were tossed in the air like the twigs of dead branches.

His deepest fear was that if floods of this intensity could rise in the dry desert, then the whole world of Humpty Dumpty was on the brink of toppling off the wall, where he had been balancing precariously for so long.

Ashok knew that, once fallen and broken, no power on earth would be able to put him back together again. Panic and conflict would impose a blackness never seen before over the whole of the human race, erasing the wonder of art and music and life itself.

There was a universal harmony in the balance of things that sustained life on earth which, just like the gossamer oat seeds that floated away in the breeze, would not be reversible once it had been fundamentally disrupted.

Ashok had the haunting premonition that humankind was entering its darkest period as nature took retribution for centuries of abuse.

He suddenly felt terribly fragile. Was this what he had been called from the desert to solve? The weight of responsibility was unbearable, but shrugging the shoulders was to abdicate the divine gift of choice.

Ashok knew that men were not gods. They had learned how to peer into the shadows of the future and correct course when events dictated. But they were not gifted with the power to see objectively and with certainty. However hard they analysed what they saw, the only certainty was ignorance. The future was another country that they were constrained to interpret through the rational eyes of self-interest and the moral eyes of virtue.

Always partial, always personal, always self-interested, always late, the rational eyes had assumed the right to direct the course of human development.

The signs had been clear for a long time that it was very late in the day to stem the tide of global warming. Flooding and heat domes and acidification of the oceans were already undermining Humpty and loosening his foundations.

Even as that happened humans were choosing to fight each other as well as nature. Wars of ambition in Europe, of restoration of greatness in North America, and of empire building in the East were prizing the cracks wider apart. Poverty, inequality and hopelessness were forcing more and more people to migrate to wherever the grass seemed greener. Walls were built to keep them out.

Was it any wonder that people in countries both rich and poor, when asked how they were, replied, "I am absolutely not alright. How did we get into this mess in the first place? This isn't what I signed up for when I joined society. There must be a more 'common good' than this. For heaven's sake, help me find it."

COMMON GOOD
WAR
HOMELESSNESS
VIOLENCE
LONELINESS
OBESITY
STARVATION

Ashok knew from his travels through time to ancient civilisations that no society had ever truly found a definitive common good.

True, there had been many great empires which had left behind enduring monuments to their glory, but their power had always proved temporary. New societies had continuously been forged by new technologies and new ideas about how people should organise and govern themselves. New empires and the idea that drove them came and went.

But finding that creative accommodation between conflicting interests, which would allow everyone to flourish according to their potential, had remained ever elusive.

Right now the world had become possessed by one new, omnipresent and all-powerful idea. The idea that the growth of 'stuff' for consumption, status and power fuelled by competition was the best operating principle to govern society. Each individual, each nation pursuing its own interests would produce the best possible outcome for all concerned.

Ashok wondered as he watched the disaster unfold whether it was the idea itself that was wrong. In the desert a stranger was welcomed and given food and shelter. There was a generally accepted duty of care. But during his travels he had seen the opposite, one man pitted against another, state against state, and now humankind against nature. Was the dominant idea itself the threat to life on earth as we know it? Was it the problem rather than the solution?

Ashok fell into a deep meditation and emptied himself of all emotion in search for the wisdom of a way forward.

He knew that historians had charted the fall and decline of empires before and after Rome. The mightiest, fabled for their strength and the glory of their monuments, had crumbled from within as people rebelled, harvests failed, people migrated, or from the aggression of more powerful forces from without.

Humpty falling off the wall was different in scale and significance to anything that had gone before. The empire of consumption knew no boundaries. It had gained dominion over the whole of humankind, the way it thought, what it valued and whether it survived.

As it conquered, the glue of time-honoured common goods of co-operation and respect for others and nature, partners in a shared journey which could not be undertaken alone, had dissolved into a race to accumulate individual wealth and national power. And the natural balance of society had been lost.

Surely a higher sense of justice, based on the shared meaning and purpose he had glimpsed at Nara, must prevail somewhere in the annals of human glory?

If so, where and what was its secret?

JUSTICE
MIGRATION
TRADE
RELIGION
MILITARY MIGHT
INVENTION
OWNERSHIP
MEANING ?
VALUES ?

Ashok turned for help to the great thinkers of previous times, particularly the idea of perfect 'forms' advanced by Plato. From his earliest days in the desert that was how he instinctively thought about beauty and truth. Eternal and absolute and unquestionable. But he knew now that society was not the same as philosophy or art, and the horrors which had been committed in the name of one absolute political good were unspeakable.

He found himself therefore spending more and more time in the company of Aristotle, who thought of the common good more as a journey rather than a destination. He reasoned that different interests competing for resources was a good thing, making the total pot of society richer because it was more diverse and resilient. Provided that is, and it was a big proviso, all parties accepted the need for continuous compromise and respect for the most sacred interests of others, to avoid the potential for tyranny that resided in democracy.

Ashok came to think of this questing constantly for balance between all things as holding the key to 'real' politics. The common good as a constant negotiation of co-operation and competition, identity and difference. Never static, and continuously capable of incorporating new technologies and social ideas. But never losing the permanent glue of enough fairness, co-operation, respect, and partnership between people and nature to hold together over time with common purpose.

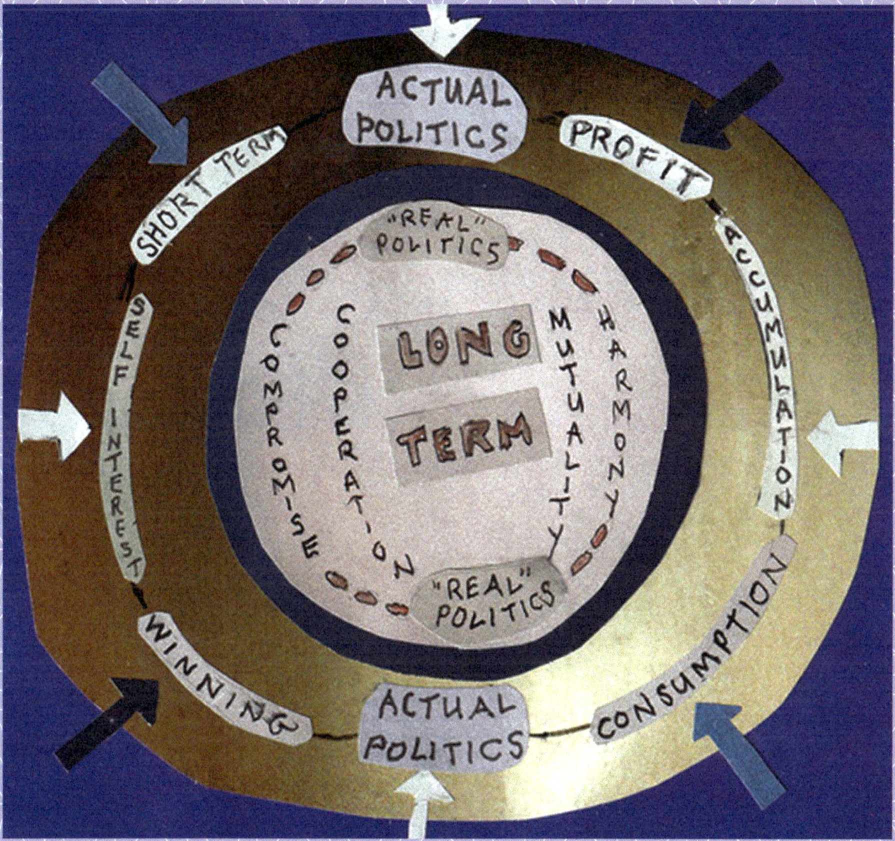
ACTUAL POLITICS
PROFIT
ACCUMULATION
CONSUMPTION
ACTUAL POLITICS
WINNING
SELF INTEREST
SHORT TERM
"REAL" POLITICS
COMPROMISE
COOPERATION
LONG TERM
MUTUALITY
HARMONY
"REAL" POLITICS

He contrasted this with 'actual' politics where the script for leaders was short-term, materialistic and competitive, putting the interests of individuals and the state first and foremost.

For him this was 'bad' power. It pre-empted the discovery of shared purpose and turned co-operative relationships based on trust into transactions and deals between competitors for gain. Once the aim was not to accept and welcome a marriage of differences but to win at all costs, people slipped easily from being colleagues to competitors in the workplace, from friends to rivals in their social affairs, and from partners to enemies in their international relationships.

Ashok saw this as 'bad' power, the very opposite of the common good, the aim at all times being to create and take advantage. Find the weakness in the other and exploit it, rather than co-operate for mutual benefit, became the measure of success underwritten by reward and development systems the world over.

This kind of 'bad' power, he thought, always had a second kind of underwrite, the threat of violence, either straightforward military might, or the much vaunted 'neutral' weight of money, to gain ascendency and stay on top.

The idea that money was 'neutral' because it allowed impersonal forces like markets to make choices, taking the 'biased' agency of humans out of decision making, underpinned the basic idea of the times. Society was 'better off', it was argued, when resources were allocated by demand not choice.

Ashok felt really uncomfortable with this idea of demand. The more he looked at money the clearer it became it was anything but neutral. If you had it, all the freedoms offered by the market were open to you. If you didn't, they weren't; tantalising promises just out of reach, temptations which bred resentment, envy, and a gradual disenchantment with work, with society and ultimately with law and order.

Money did some other things too. It was the measure of growth and therefore the driver of pollution. By excluding those who couldn't pay it took away the feeling of belonging to something bigger than oneself and willingly committing to improving productivity. It starved public goods in favour of private consumption. And, worst of all, it voided society of the shared meaning which is the ultimate glue of society.

When Ashok looked at it like this, he wondered how such a system had come to dominate the way people thought about how society should be organised, and why it was tolerated.

These were complex questions. Ashok knew that changing the dominant story and the values that modern society lived by was a gigantic task, a form of 'new world building'. He also knew it would be no easy task to get those with power, who benefitted from how things currently were, to relinquish authority and make a new story work.

He was by now streetwise enough to know that whatever might emerge as the new guiding principle of how people lived together in society, governments would not be the agency of change. For a new story to have a serious chance of changing behaviour it would have to be co-created, co-produced and co-owned.

For inspiration he thought about how financial markets 'discovered' the price of things by putting all the information about supply and demand on the table, and decided the best way of proceeding was to copy the markets and create the possibility for different interests to 'discover' a common ground through a no-holds barred encounter.

He would therefore take the great risk of convening a Conference of All the Parties in a spirit of discovery and conciliation.

Using the exceptional powers of the amulet to enter the minds of anyone he wished, he gathered the most powerful military leaders and the chief masters of the financial universe and set them down in the desert, together with a selection of the 'governed' including: the successful, the ordinary, the unheard, the unwashed and the disadvantaged.

Ashok was interested to see how each of the Parties used the space in the desert theatre he had created for them, which they felt most suited their place in life.

The Military Men set up a dais at the far end of the maidan, which gave them a clear line of sight and, if necessary, a clear line of fire to control proceedings. They sat in a tight, grim formation, a phalanx of authority exuding power.

The Masters of the Financial Universe congregated at the other end, noisily exchanging greetings of goodwill, whilst at the same time carefully withholding information which might afford an advantage, and sporting their feathers to each other like cocks at a market.

The unorganised 'Others' sat wherever they found themselves, randomly distributed about the floor, some sitting back to back for support to ease the discomfort from long journeys on foot or by camel, others stretched out asleep in hollows dug out of the sand.

Ashok was struck that each of the Parties consciously ignored the others as not being 'part of them', and waited impatiently for Ashok to tell them what it was all about.

Ashok set the agenda by teleporting pictures of the mess the world was in, and how close civilisation was to self-destruction, embedding the big issues of global warming, gross inequality and the permanent threat of military conflict into the minds of all of the participants.

He used Deep Artificial Intelligence to trigger a range of scenarios including dramatic images of whole peoples fleeing poverty, soil erosion, water wars, universal obesity, conflict, terror, alienation and the absence of hope.

After many days of debate each of the Parties gradually came to accept the possibility of impending catastrophe but differed radically about why it was happening and what to do about it.

The military men and the money brokers argued the system in place, with all is disadvantages, was the best and only feasible route forward to get the growth that made the world go round. They reserved their harshest criticism for solutions that changed the way companies were governed and were deeply sceptical of actions designed to reduce pollution and waste. They argued it was right that the existing system buttered their bread most sweetly, as they were the creators of the wealth that benefitted everyone.

PROFIT
climate change
nuclear wars
soil erosion
civil conflict
migration
terror
absence of hope
pollution
political agitation
RAGE
VISION
SOLUTION
RISK

The rage of the great majority in response was palpable. For the first time ever they were able to express their feelings directly in an open forum to the leaders they held accountable for the way things were.

They rose as one from the desert floor demanding action on the issues of poverty, war and climate disruption which impoverished their lives, with hunger and hopelessness stalking their every day.

Ashok watched the growing fury anxiously. There were no signs of compromise or conciliation, and he knew it would take only the smallest spark to ignite a bonfire of the furies on all sides.

His search for a common good approach to avoid the world collapsing into a wasteland of violence and emptiness looked as though it could precipitate the very disaster he was trying to avoid.

Increasingly desperate to find a way to change the mood of the encounter, he invited representatives of the 'unpowerful' to come forward and tell their own stories.

The first to come forward was a very poor housewife called Ada who came from a very rich country. She recalled the fear from nightly bombing raids over England during the Second World War, and the cabbage water and potatoes which were the staples of her diet for six long years of wartime rationing. Even more strongly, however, she could still feel the sense of belonging and shared purpose in the small community where she lived.

Then one day when the war had ended a man came to her front door offering a bag of money to buy her council house from her. "You could make a big profit," he said, "and never again have to have your gate painted the same tedious colour blue as the house next door".

Faced with the opportunity to be suddenly rich, she replied in a way that took him completely by surprise.

"You are standing on my front step which I have scrubbed with pride every morning for the last 50 years, and you tell me I can have money from a magic tree to buy something which is already mine. This is my home, not something you can buy and sell for profit. There are more important things in life than the colour of my front gate. You insult me, get off my step."

"And anyway I like the colour blue."

When she sat down ragged men and women howled their approval.

A young man then pushed himself forward and said he thought the woman spoke nonsense. Surely given the opportunity to make money, you took it when it came.

He was born poor but had made a fortune buying and selling houses. He holidayed at Val d'Isere, and ate at the Coq d'Or, which was as close as he wanted to get to the common good.

The woman rose once more to her feet and asked him to look inside himself for a moment and ask the child he had once been whether he was proud of the man he had become.

Ashok willed him to be truthful and, as if in a trance, he recalled the time his mother who lived in poverty, to his intense surprise, had felt insulted when he gave her a Christmas tree decorated with bank notes instead of holly. He had felt crushed and ashamed by her rejection, realising he had allowed his love of money to intrude on the sacredness of Christmas. His intention had been good, but he knew deep inside that the child he had been would not be proud of the man he had become at that time.

Now, once again, he felt crushed and ashamed by the emptiness of the life he had described.

Ashok then invited a Native American Chief to speak about a more distant and ancient culture.

Chief Whitehorse said, "My people since the first sun rose in the sky believed nature was a gift of God, not something that could be privately held and turned into money for self-advantage. Hoarding and accumulating wealth was considered shameful, allowing one man to become richer than another and destroying their equality of being. Our values were preserved by living closely together in the Smoke House which set clear boundaries of behaviour for the young. We did not need policemen to enforce them. The old were honoured for their wisdom in showing us the way. We were guests of the spirits which owned the forests and the rivers."

"Now," he said, "these values were changing. The old world of co-operation and respect was being lost and money was becoming the master of mankind."

He was listened to in silence and when he sat down, the crowd growled their support for the lost way of living he felt so deeply.

Finally Ashok invited a yak herdsman from Nepal to come forward and close the conference.

Pemba told of his life in a remote village in the high Himalayas where barter and co-operation were the norm. By working hard and pooling the milk of his own yaks with others, he had been able to negotiate a better price and gradually grew richer than his neighbours. Diligent by nature he had used his profits to build the biggest house in the village and surrounded it with a large wall to protect his yaks from predators.

At first his success made him happy but gradually he was conscious of a change in his relationships with his neighbours. They still made 'namaste' when they passed in the street but he was no longer included in their daily routines and conversations in the square. He began to feel like an outsider in his own village, no longer able to feel the contentment of belonging at the end of a long day which came from a sense of 'coming home'.

"The price of my success," he concluded, "was unbearably high. I became a stranger in my own land."

As he had done so many times before, Ashok jotted down the ideas and connections that came into his head as he listened to the stories that were told. He scrawled on the wall of his hut what he had heard:

- some things are more precious than money
- homes, communities and relationships are sacred
- money is not love and can demean both the giver and the receiver
- nature is a partner with its own rights in discovering the public good
- excessive inequality kills the spirit of difference and acceptance
- shared purpose is the glue of society
- the common good is a process not a final state.

Before sitting down he asked everyone to think hard about one final question. "These wisdoms we have co-created and co-produced together," he said. "Do we really own them as the principles by which we wish to live?"

• many things are sacred and should be protected from money
• homes, communities and relationships are three of those things
• money is not love and can demean both the giver and the receiver
• nature is a partner with rights in the public good
• excessive inequality kills the spirit of co-operation
• shared purpose is the sacred glue of society
• the common good is a process of negotiated discovery not a final state

When it came down to it, although the Military Men and the Financial Masters of the Universe had been party to all that went before, when they went away and thought about it together, they weren't at all keen on the implications for their own positions of power and wealth. In no time at all they had formed up behind a ring of steel and weaponry and rejected the principles out of hand. The Others, who were much more numerous, became increasingly angry and although they were unarmed the potential for a violent confrontation was ominous.

Ashok was distraught. His experiment at 'new world building' by redirecting the journey of mankind from competition between men and its onslaught on nature, without at the same time curbing the divine gift of creativity and discovery, had failed.

If he had any doubts beforehand, it was now abundantly clear it would take much more than a shared story, carefully co-produced using shared information, to challenge the invisible idea that had such a grip on man's thinking.

The idea that unencumbered self-interest is the elixir of human well-being was obviously too powerful to change by reason alone. Something else was needed.

Shattered and exhausted, he decided he had to close the conference before disagreement turned into violence.

He had failed in his quest to create a common cause, and the only option he could see was to bury the scrolls on which they had inscribed their wisdom deep in the desert, in the hope they would be rediscovered one distant day in the future, rather like the Dead Sea Scrolls. A day when humans were truly ready to take responsibility for each other and the planet that sustained them as a home that provided them with their needs, and not as a lodging house to be mined and traded for profit.

Little did anyone know the day of reckoning was already upon them.

ASHOK WATCHES HUMPTY FALL
AND LEADS A CRUSADE OF REGENERATION
NEW SCHOOLS OF THE COMMON GOOD

Ashok watched in despair as for seven days and nights the arguments raged at his Conference in the desert. The attempt to find a consensus to stop the world toppling like Humpty into a black hole of environmental catastrophe and social violence was in tatters. The gulf between interests had only become wider, and fights had broken out as tempers rose.

The men in possession of the old way, the Military Men and the Masters of the Financial Universe, refused to compromise. The world of free markets and maximum competition was the only way to go for everyone's benefit, rich and poor alike.

The rest, the Concerned, the Poor and the Afflicted looked at each other in bewilderment and increasing anger. They could see the spectacular benefits of that world for the few, but they could also see the coming environmental wasteland for everybody, pushing humanity into unknowable territory. They argued that a new world had to be born that was more co-operative, more fair, and more sustainable, one working above all in partnership with nature.

In desperation Ashok called timeout to allow tempers to cool and totally exhausted, sank into his own troubled sleep. But not everyone slept.

When the camp woke at first light it was to find 13 bodies hanging from the giant fig tree which overhung the maidan. Those who had argued most strongly for change – Ada who refused to trade her home for money, the boy who had tried in vain to buy the spirit of Christmas for his parents, the Chief who lamented the loss of community, and the yak herder who lamented his exchange of relationships for commercial success – had been arrested and executed in the night for incitement to treason.

Unable to believe what he saw before him, Ashok knelt in the sand and wept. The sun rose relentlessly in the sky, regardless of the human saga being played out below. Ashok remained prostrate throughout the great heat of the day, shattered at how quickly differences turned to violence and how hard it was to compromise, even as the world rushed headlong towards disaster.

He could discern no pattern of hope in the horror of what had happened. It seemed there could be no shared way forward until the pain from the present pathway became unbearable.

When he finally raised himself as the evening shadows fell, he saw that he had not been alone in his vigil. Everyone, the sick and disadvantaged, as well as the Masters and the Military men, were kneeling around him, quietly sobbing in homage to the dead. Raising his hands in blessing he proclaimed each of the dead a Martyr of the Desert Scrolls.

Then, turning towards the silent crowd, he proclaimed each and every one there a Messenger of the Common Good. When they realised they too were included, despite what they had done, the Masters and the Military Men felt the enormity of the honour and the burden of their shame intensely, throwing themselves once again on the ground in repentance.

Ashok raised them up, and singled them out for particular praise. Then, with a heavy heart he explained there was no choice but for everyone to go back to the places from where they came, there to await the call to return and build the new world when the time was right. Reluctantly and saddened by their failure, they left for homelands far away.

MARTYRS of the DESERT SCROLLS

As the Messengers returned they became wide-eyed in disbelief at the devastation they passed through. During their time of doubt in the desert, the planet had taken matters into its own hands. Great pieces of Humpty had broken away and fallen, amidst scenes of unimaginable horror and devastation.

Those travelling to Europe trudged through a wide corridor of towns and villages laid waste by war between Russia and NATO. A few isolated figures moved amongst the smouldering landscape, aliens in protective masks and clothing.

The Indians travelled into a heat dome so intense every village they passed through was silent with the torpor of the dead and dying.

In China and North America, vast regions were covered by floods, a watery wasteland of the drowned and the sick from unknown diseases.

Africa was aflame with a thousand civil wars and the flow of migrants seeking a better life North had been forced back on itself in the face of new, unbridled savagery.

Democratic governments everywhere had been replaced by the military.

Ashok watched his followers disappear with a heavy heart, feeling a great weight of personal responsibility for what had happened at his Conference. In the middle of a restless night he woke with a violent start, as if a torrent of invisible horror was flowing back to him from the Messengers. Vividly in his dream he had felt the desperation of the whole of humanity teetering on the brink of a black hole from which there could be no recovery.

He could see the old order shifting on its axis, but the shift was mindless. There was no form or purpose in what was happening, only chaos. Amidst the wailing cries for help, the only response was the sound of doors being hastily shut and walls urgently constructed to ward off the hopelessness and seething violence which was all pervasive.

Ashok could feel himself falling helplessly into the dreadful darkness of the void into which the world itself was falling, fearful at the horror of what he saw, powerless and now doubly guilty that he had not been able to do more.

For comfort he fingered the amulet which had been his constant companion on his journey out of the desert and found himself flying once again through time and space to seek advice from Gaia, his all-seeing and all-caring Mother.

Unconsolably, he told her of the tragedy that had happened in the desert, about the killings and how people had become disconnected from the love of each other, seeking only advantage for themselves. Even the most far-sighted and well intentioned felt powerless to change the pathway they had accepted for so long. They knew in their hearts they had willingly organised themselves to compete for advantage, one man against the next, measuring their well-being and purpose in life by the accumulation of money. Slowly, and then rapidly, the glue of relationships which had bound them together in association had dissolved.

It was the same between countries. The solidarity of peaceful co-existence had given way to competition through war. Compromise and respect for differences had been dismissed as the way of the weak. Leadership was now about negotiating deals from strength, pursuing self-interest reinforced by threats.

After a long silence Gaia asked him whose decisions and choices were these that were so disconnected from love and brought such destruction? "Remember," she said, "power can only ever be borrowed. It does not belong permanently to one idea. If power is usurped and abused by Governments and their institutions, they must be helped to see a new idea of society which they can follow."

"The idea of giving rather than taking," she said, "resides in every heart, but it is not always easy to find." Looking at him with softness in her own heart she said that, dejected and exhausted as he was, it was his duty to start again. "Reconvene your Messengers and discover together the form of that new idea. Find their wisdom, and you will find the way."

Ashok was instantly rejuvenated by the sincerity and simplicity of his mother's advice and sent messages through the ether to his Messengers to turn about and return to him as quickly as they could.

Weary and weighed down by their own desolation as they trudged through a blighted landscape, they were startled to hear his voice calling them to turn back. But as one they obeyed and began to retrace their steps, their mood light again with hope. Now they marched with a new vigour, talking excitedly and singing with joy.

Hearing their songs and feeling their pride and hope amidst their own desolation, people along the way came out of hiding, cautiously at first, then enthusiastically, falling in behind the Messengers in huge crowds and swelling their ranks a thousand fold.

Among them were bee-keepers, teachers, mothers, the lame and the infirm, the rich, fishermen and shopkeepers, the wise and the not so wise. Individuals and members of every kind of association were lifted up by the message of hope and became a joyous multitude.

Despite the grim trek they made through devastated villages, the travellers felt uplifted by the purposeful joy of their companions in the face of the catastrophe which had overwhelmed them all, and strong in the resolve to live their lives by a new, more co-operative story.

ASHOK WATCHES HUMPTY FALL AND
LEADS A CRUSADE OF REGENERATION

When, after many days of marching, they arrived back to the place of their conference in the desert they were welcomed by a joyous Ashok, who immediately tasked them to dig out and recover the scrolls they had carefully hidden in the vast expanse of sand surrounding his dwelling.

Despite the horrors they had encountered as they passed through devastated villages and landscapes, they felt reenergised and immediately set to with a will, working hard in the extreme heat, co-operating as one with shared purpose without leadership or promise of reward.

Shouts of great jubilation arose as, one by one, the scrolls were uncovered and borne aloft by many willing hands. Triumphantly they were brought back to the maidan, where the Messengers and their new multitude of followers, who had not been party to their creation, had assembled in vast numbers without end.

Ashok received each of the scrolls reverently and, as more were brought forward, a sense of awe and excitement grew at the wisdom waiting to be revealed.

The assembled followers felt like the prophets in ancient times must have felt, bearing the keys to renewing a world torn apart by division and already tumbling over the brink of devastation. They had seen with their own eyes the horrors that brought, and couldn't suppress a great fear alongside their newfound euphoria.

Ashok also felt that fear and in his welcome stressed there was no time to spare. The world was already collapsing. They had witnessed in their journeys home the catastrophe unfolding everywhere and accepted instinctively the urgency of turning the scrolls into new values and behaviours in the minds and lives of everyone they had made contact with.

They listened and accepted with joy the heavy task he placed upon them to promulgate and carry forth within their own lives the conviction that a new world could be built, one which 'would cultivate a new collective mind, joined by the common threads which brought lives together. One which would live out the common ground of everything in its daily existence'.

As he called for the scrolls to be brought forward he emphasised that the collective mind could be good only when it honoured and sustained differences between peoples, and enhanced the well-being of all in a spirit of conciliation and co-operation.

SHARED PURPOSE

The first Scroll that was brought forth read simply 'money can't be eaten'.

Immediately emotions ran high. Money, for those far from the breadline, was exactly that, something to be indulged and eaten. The more of it they had, the more choice they had to satisfy their appetites for new things. And the more sharply they could distinguish themselves from their neighbours next door. Surely they argued these were vital parts of the good life?

The Masters of the Universe and the Military Men who, because of their shame took their roles as Messengers with extra responsibility, supported those arguments. How would we know, they asked, the value of something except by its price in money? And what fairer way to set a price than what consumers are prepared to pay? This was surely the way the world worked best, the way of freedom from Governments and from biased human choice?

But, little by little, different, smaller voices began to make themselves heard. A follower from Mecca said that true freedom, the freedom of the soul, was indeed unbounded and not something that could be imposed but unbounded hunger and the absence of hope were not freedoms either. They were killing people here and now. Market prices were not free prices except for the rich. Different voices pitched in, arguing that other basic things were like food, and should not be available only to the rich.

Many argued that water, education and security were eternal goods with intrinsic value, not something to be bought and sold exclusively by those with money.

MONEY CAN'T BE EATEN
VALUE IS AN INNER QUALITY
VALUE IS SOMETHING INTRINSIC
VALUE IS A FREEDOM
EDUCATION
DEVELOPS
IMAGINATION
AND VALUES

By the end of the seventh night the debate had extended beyond basic rights to education, homes and the role of business.

The 'idea' of education to prepare everyone for work was quickly accepted as a right. But voices argued for a yet higher duty than that. They said that imagination in art, connectivity between ideas and civic awareness and co-operation were the cornerstones of an open society, and therefore core to the larger idea of education.

Similarly with houses. As bricks and mortar they were obviously financial assets, but as homes with inner lives and memories, they had an altogether higher level of being with intrinsic value. They could never be just transactions with a financial price determined by markets.

Business, it was quickly agreed, you couldn't do without, since it was the vehicle for getting things done. How could you criticise something that touched every aspect of life? But when a woman slowly raised herself from the floor and stood silently with hands growing from her shoulders, the mood changed. People were reduced to tears at the sight of what can happen when the animal spirits of business go wrong. Others followed her lead and stood to tell of lives lost as buildings proclaimed safe became infernos from which there was no escape, flames accelerated by their protective cladding.

The distinction began to form of business for profit and 'fair' business with a wider purpose. The duty of all business, it was argued, must be first and foremost to do no harm and add value, either to people or the environment. Then and only then should its objective be to make profit.

DECEIT
AGGRESSION
CORRUPTION

As they listened the Military Men and the Masters began to look at education, homes, companies, and infrastructure through new eyes. The idea took shape of a common treasury of shared goods requiring consistent care, repair and creativity for both individual well-being and for society as a whole. They began to ask themselves how profit and love could learn to lie down together for mutual benefit. Surely humans were not preordained like the tiger to leap regardless of the consequences?

Even so, they were not yet ready to accept that the very axis of society needed to change from individual profit seeking and competition to co-operation and love between people and with nature. Just when there seemed no way forward, from a far corner a child's voice could be heard asking, "Did I really lose my mummy, my brothers, my sisters and my friends in the flood because we all loved each other too much?"

It was as if a charge went through the entire assembly like a bolt of lightning. Her question reminded them that they had already lived the old model of accumulation and advantage. It was the source of their problems, not the answer. The results were visible all around them. It was not love and co-operation which had destroyed communities, bred corruption in public office, sparked war between countries, and elevated the pursuit of growth and short-term profit by companies and governments.

It was the endless quest for more. And just a short step from there to the destruction of the environment they had just journeyed through. And of life itself.

They could see these things happening in the twilight around them as Humpty began to fall and break. And, as one, they demanded a new story to live by that gave them purpose and hope beyond money.

PROFIT
LOVE

The second Scroll was held high by a woman with impeccable credentials for the modern world of growth and individual rights.

She had a double first in economics from Oxford, best in class at Harvard, youngest partner in a global law firm, fluency in the languages of the new world Mandarin and Arabic. She had become a trusted conduit of capital flows for the oil rich. And she was a woman.

Having just fought its way through the idea of education being a basic right, the multitude seethed at this flaunting of the symbols of power and conspicuous consumption.

Their instinct was to reject and they hissed and booed as she talked.

They were still only half listening when she told of meeting a child in a temple in Bhutan, clad only in rags and covered in sores, kneeling on the cold stone floor.

Their rage was such they missed the point when she told how she offered her scarf for the child to kneel on, but the holy man intervened and said, “she does not need it, she kneels already on the pillow of gratitude.”

ASHOK WATCHES HUMPTY FALL AND
LEADS A CRUSADE OF REGENERATION

But there was complete silence when she told how she felt so humbled by this experience she had renounced her previous life on the spot.

She had experienced at that very moment an epiphany, which she said, had turned her sense of purpose and life upside down.

It was like "a joyful falling away of the staleness and alienation of my previous life, rich in things but without common ground or humility."

The crowd now listened to her with rapture. With those few words she had become an angel of light instead of an emblem of selfishness.

They roared their approval when she described the mission of her life now as "a search for the way of gratitude for kindnesses so small they could not be seen by those on the motorway of progress."

The Scroll she held high was inscribed with Maya Angelou's words 'let gratitude be the pillow upon which you kneel to say your nightly prayer'.

let gratitude be
the pillow on which
you Kneel to say
your nightly prayer

The third group invited to tell their story were an indigenous tribe from Colombia. They too spoke about motorways, a motorway of so-called progress cutting its way through their ancestral forests. The loggers argued they had the legal rights to the timber because they had bought permits from a government short of hard currency to finance services such as education and public health.

The tribes did not believe these rights belonged to the government in the first place. Even if they did they were sacred rights which could not be bought with money.

The trees were members of the forest community. They had spiritual connections to a communal past and a shared identity with the people who lived in the forest. They were part of a system that sustained life and ensured the balance of the winds and the rains in the interests of the entire world.

"The tribes," they said, "had an unbreakable responsibility to preserve them as the stewards of this common ground. The currency of that stewardship was partnership with the natural world, not money."

The scroll they raised read 'forests too need love'.

forests too need food

The fourth group was a mixture of hermits, businessmen, academics, the rich and the very poor. Their spokesperson was a Jain, who said his spiritual goal was to achieve the all-knowing state of 'moksha', whose earthly expression is abstinence.

For many generations they had maintained a constant state of moral rectitude, strictly observing the rights of every form of life to live in peace. They believed that doing no wrong was the highest form of spiritual and social awareness.

At the same time they had accumulated vast wealth over many generations.

The younger members of the family increasingly could not live with what they saw as a separation of moral responsibility and material advantage. They felt they had lived a double life which they now repented.

The family of four generations had decided together therefore to make recompense for their blindness to the unsustainability of their obsession with wealth. In a communal act of penance he, his sons, his father and his father before him had renounced their total worldly possessions to live a life of unrelenting austerity. He described how they felt released and joyful in their newfound simplicity and poverty.

They had pondered long whether their epiphany was unique to them and spent much time searching through the wisdom of great thinkers from the past. The views of Rousseau and von Herder, formed amidst the wealth of industrial Europe, led them to conclude that they were neither alone nor the first to believe that material progress led to moral degeneration.

This corresponded to his own experience that the values of the Enlightenment of reason, objectivity and science had led to an excessive individualism which had usurped the values of faith, place, relationships and community to a point where the organic foundations of society had dissolved.

"The logical conclusion," he said, "was the growth obsessed society we lived in today, where a few men were able to capture as much of the world's assets as the poorest half of the global population combined. And because of their wealth not their virtue, were able to exercise huge political influence even in the most democratic societies. His family had come to the difficult conclusion that such a society was fundamentally diseased and out of control."

The enemy, as they saw it, was no longer the old one of competing ideologies, where a personal choice could be made. It was the ubiquitous system of concentrated private capital facing off with an impoverished multitude, where choice was restricted to financial ability to pay. The evidence that society without effective mediation was not only morally bankrupt but also unsustainable, was all around them in plain sight.

The scroll he held high said 'individualism is barren of duty'.

individualism is
barren of duty

As evening fell the last Scroll was paraded for all to see. It was held high by a smallish white man dressed, despite the heat, in a beanie hat and jeans. He explained he had been advisor to Prime Ministers on 'why States fail'. His scroll read simply 'failure starts at the top'.

In a provocative speech he dismissed the Military Men and the Masters of the Universe along with Politicians and the Institutions that supported them. He described them as lackeys in a system which had overwhelmed them. The combined forces of authority, money and political parties had lost the art and power of true leadership, reacting to events and reduced to representing their own partial interests.

The most urgent issues of social cohesion, productivity, migration, global inequality, smartphone addiction, control of social media, loss of community and relationships, and most critically global warming and the destruction of rivers and forests, languished in the narcissism of their minor differences.

He argued the present system equated electors with consumers, competing with empty promises of future growth and sunny uplands. When the event horizon is the next election, vision is narrowed down to 'work' which is so much easier than thinking. High achiever stupidity is valued over a critical mind. The skills of public leadership are reduced to presentation and processing power. Galley slave education drives the galley faster, but always in the same direction. Public standards collapse and accountability exits through the revolving door en route to the next senior appointment.

The very system of governance was creating the conditions for the real issues to fester and pre-empting the emergence of a common good.

failure starts at the top
social co-hesion
CLEAN WATER
INTEGRITY
PUBLIC STANDARDS COLLAPSE
productivity
TRUST
UP
accountability

This was revolutionary stuff he was proclaiming. It was listened to in a near silence, which was difficult to interpret. Despite the passion and sincerity with which it was delivered, Ashok feared it had simply passed over the heads of most people.

But it hadn't.

Up to this point the followers had been intensely engaged as the individual scrolls were paraded, and the significance and implications of each was hotly debated this way and that. They may have been mostly simple men from ordinary walks of life but they had all directly experienced the horrors of a world in the first stages of falling apart. And they didn't like what they saw. Deeply committed to finding a new way forward, they were struck by the wisdom of what they had heard.

But so far, despite the common threads that had emerged, they couldn't see the story which joined them up. Suddenly now it was clear how interconnected the issues were with each other, and how it was the system of values and governance which prevailed which was driving the unsustainability of an economic and social system based on competition and individual rights.

Ashok could feel a deep shift in understanding and mood, like a seismic shock reverberating underground, as the significance of what they had heard sunk in.

A sick society produces sick members and sick results.

state failure
money is not value
the economic system is BANKRUPT
let gratitude be the pillow
preserving trees is not CHARITY

The realisation was overwhelming. The entire system, which had worked for 300 years, had to be regenerated from scratch. What had become known as the Industrial Revolution and the Enlightenment were truly wondrous in their time. Science and courage at sea brought phenomenal wealth to a small buccaneering group prepared to risk their lives scouring the oceans for trade, and even greater riches to a more sedentary group prepared to risk their capital in unpredictable ventures to far-off places. Enterprise and capital had been the building blocks of modern economies.

Since then growth had become the overriding priority of Governments. Year after year production and extraction from nature dominated human activity. Perpetual growth had pollution baked into it. So much so that the system, its values and its practices were now tipping the entire environment towards disaster. Rivers, seas and air so polluted the Earth's temperature was getting hotter and hotter, its forests, flowers, birds, bees were being depleted, major cities across the world would be submerged as ocean levels rose, and the environment that humankind depended on for survival was being laid to waste.

That much was known. What hit home hardest was that the minds and values of men and women the world over had become blind-sided to the consequences of what was happening by the obsession with personal gain and commercial advantage. Human consciousness had become corrupted by its own experience.

The realisation ran like a wave through the crowd. They could suddenly see that a sick society inevitably produces sick bodies and sick minds. And governance that prides itself on dealing with events after the fact, is guilty of failing in its primary duty of well-being and the nurturing of hope.

As the camp slept that night Ashok thought back to his first venture as a dustman in ancient Ur. His fresh eyes from the desert had found a way to reduce the waste that was overwhelming the city, but it was the citizens themselves who had shown the will to permanently solve the problem by changing the way they thought about consumption and waste.

For the first time, as he slept, he felt the excitement of seeing that the 'Way' and the 'Will' had to go together, and neither was enough on its own. He knew the issues raised in the desert were much, much larger than the single issue of waste, but was it possible the same principles applied?

Putting sticking plasters over deep wounds could only hide them for a time not heal them. The key had to be in the way society chose to organise itself. The present genie of individualism and obsession with private wealth accumulation had got so far out of the bottle that the idea of society as a living, interconnected whole had been lost.

Nothing less than the collective mind of humankind needed to change. Was that the challenge he had been set?

As he meditated Ashok saw that the 'Will' would only come from a new Settlement of Understanding that society was for mutual benefit. Governments had become part of the problem. Power to effect deep change lay within the multitude of interests which make up society. They needed a new story to live by. One which galvanised their understanding that true richness comes from reconciling not supressing differences. But they needed more than a new story. They needed a new 'Way' of governance, based on refreshed institutions at all levels, working openly to mandates of co-operation.

BUSINESS
CIVIC SOCIETY
STATE
labour
energy
nature
love
balance
legitimacy
consensus
purpose
BIN OF PAST HOPES
No will
No way

Ashok felt at that moment a deep calmness descend on him in the conviction that both the Will and the Way could be found to stop the world from toppling irreversibly into an environmental wasteland, filled with its own fury, and that the Messengers and the Followers were the chosen vehicle to make that happen.

As the first light of dawn showed in the sky, Ashok told the vast throng assembled before him about the vision which had come to him that night, a vision of a crusade to regenerate the collapsing world around them. A crusade that would not be led by existing political systems and their institutions. They were the brakes holding the present system in lockdown.

For a new world to be born, new values of co-operation, respect, partnership and love would have to come from the people themselves. New institutions with refreshed mandates would have to be fashioned in the spirit of the common good. New leaders had to be discovered from the multitude of networks and concerned individuals who could see the imperative for change but felt presently impotent.

"It was," he said, "individuals of extraordinary fortitude and courage who would go forth and build this new world, place by place, piece by piece."

GO TELL

The Followers listened in awe, but asked themselves where were these extraordinary Teachers to be found. Ashok told them not to look for strangers in some distant academy or distinguished office far away. "They are already with us, here, present now" he said. "It is you who will be the Messengers to carry this story to the ends of the Earth.

"Remember when you came to this place you had great fear in your hearts. The divisions between you were such that taking life was easier than compromise. But you have found comfort and wisdom in each other. You have learned the art of conciliation, the truth that together is better than apart. You have the energy, the understanding, the commitment to building a new world.

"No one could be better equipped than you to lead the revolution which must happen. And it is a revolution. Your duty is not to instigate violence in pursuit of the common good but to practice the art of conciliation in everything you do.

"Put yourselves at the service of every community you settle amongst. Act not as strangers, but as members and teachers, living and sharing the same concerns. Listen attentively to the needs and hopes of the people in each place, and instil by example the values and relationships we have constructed together here in the desert.

You will be the 'Way' and the 'Will' of the new world."

LISTEN
CONNECT
TEACH
CONCILIATE
VALUES
NEEDS
HOPES
RELATIONSHIPS
GO TELL THEM

He warned them they would not always be welcome. The forces of the old way would resist and attack. They would often go hungry and their lives would be hard. They would suffer pain, abuse and imprisonment. They would be tempted by offers of comfort and high position to stray from the path they knew to be right. Some would succumb but in their hearts they would be changed forever.

The prize would not be material gain nor the trappings of high office. The glory would lie quietly in the knowledge that people were returning from their exile where they had been lost in the old paradigm of power and advantage. It was in the hearts and minds of those who return, who will be legion, that the Messengers and the Followers would be celebrated for showing the 'way of right values and connectivity' out of the wilderness which prevails.

When he sat down, there was a long period of silence while the enormity of what they had heard sunk in. And then, as one, the multitude rose in salute and immediately prepared to set forth once again, this time on their extraordinary journey to build a new world.

They departed in a pandemonium of excitement and shared purpose, their ranks swelling as they went along, small groups breaking away here and there to found the new schools of the common good, instantly replaced by newcomers anxious to share the joy and promise they witnessed.

GO TELL THEM IN SCHOOLS
NEW SCHOOLS OF THE COMMON GOOD
Imagination School
Sustainability school
co-operation school
CONSCIOUSNESS school
Trust School
$f(x) = \frac{1}{dx}$
Santa monica BLVD
Hollywood Blvd
margo's HOUSE
SOME BAR

At first the Messengers worried that they had no destination, simply stopping wherever they found a problem. Sometimes they were not welcome. But they looked always for ways to be helpful and, bit by bit, they established trust by finding ways to solve problems. And so it was, from small beginnings, the threads of the common good came together, knot by knot, into a necklace of shared purpose.

Ashok's original encounter in Ur with waste and pollution had been the start of his long personal journey. The Messengers found the same problems were common in all the communities they passed through. People resented the toxic fumes from cars and lorries which left a bitter taste in their mouths from sunrise to sunset. They closed their windows in a vain attempt to keep out the smoke from the power stations and manufacturing plants, which was so thick on bad days they could barely see their hand in front of their face. And everywhere there were too many people for the houses and hospital beds and schools available.

Growth had replaced poverty as the source of distress. Water sources were drying up as temperatures rose, and water wars were forcing whole villages and regions to relocate as their crops failed. Fuel for cooking what little they had was scarce. People were hungry. Everywhere they were fearful of the State and its military. Border villages had to be ready to flee at a moment's notice. Towns were divided by religion but thickly connected by fumes and vehicles. Inequality was overwhelming and beggars sat in front of the marble facades of new office buildings. The few who succeeded in finding the stairway to the stars kept the old values and installed doors behind them as they climbed.

sharing water
consumption
THE NECKLACE OF SHARED PURPOSE AND RIGHT VALUES
fumes/
vehicles
waste
indulgence
inequality

The lives of the Messengers were indeed hard. Finding the common good in the midst of so much suffering was not easy but they found strength in living the values which they preached. Little by little they became respected for their wisdom and were sought out at the end of the day, as Ashok had been in his hut in the desert. Gradually the evening gatherings became more like schools, bringing imagination and practical solutions together on issues which were current and urgent. Issues like community organising, accountability, conciliation, mutual economics, waste and consciousness, issues which sprang out of the daily lives of ordinary people, for which there were no ready-made solutions.

And in their darkest hours they knew they were not alone, certain that Ashok would always be present in their lives when the need was greatest.

On that final day, and when the last person had departed, Ashok had lain down exhausted in the shade of the fig tree and slept for seven full days and seven nights. His dreams were filled with the memories of those who had sacrificed their lives in the very place where he now slept to find a new way of being together. He was deeply troubled at the complexity of regenerating an exhausted and violent world but certain that something momentous had started in this very spot. Something which, although the journey would be long, would change the world into a better place.

While he slept, his mother, Gaia, watched over him from her place in the sky. She could see the torment in his soul but smiled at his resolution.

Despite the darkness which was coming, she could see the idea of the common good, dormant but alive within the darkness. She knew that Ashok was its Teacher and that his followers, with his constant care, would inspire the 'Will' and find the 'Way' to put Humpty together again.

AN EPILOGUE OF ENCOUNTERS WITH ASHOK IN PRESENT TIMES

The following stories are about people who have found ways to get off the bus and encounter the world in a new way.

They are real life encounters in present times which took place in Bulgaria, Lithuania, Ireland, India, Alaska, London, Wales, the Sahara, Bangladesh, Erewhon, and the Kalahari sparked by the teachings of Ashok.

In their accumulation they suggest a new way of being.

INTRODUCTION

It was Ashok's birthday, 5,000 years to the day since he heard a voice calling him from the night sky to leave his place in the desert and set out on a long journey. He was not told where to go or why but remembered vividly arriving at the great city of Ur on the banks of the Euphrates.

He had been astonished at the contrast between its beauty and dynamism, and its despair and desolation at the plague and pollution caused by the rotting refuse in its streets. On this special day he thought back fondly to the co-operation and good luck he had in sorting that problem out when he was appointed the First Dustman of the World, and his subsequent veneration as The Wise Man of the Desert.

Although still a child in cosmic terms he felt the earthly weight of responsibility for finding a new way of being together in society, not based on violence, competition and material excess. He was acutely aware that key Followers had suffered terrible deaths and relentless persecution as part of that journey.

Gaia, his mother, could sense his weariness. To cheer him up she had invited the various Schools of the Common Good he had established around the world to send him birthday greetings. She had asked for stories about how, despite the wasteland which war and growth had created over vast tracts of the globe, there were examples of extraordinary people doing extraordinary things to make life on earth better and more noble.

These are some of those stories. They tell of real-life encounters in different parts of the world by people of all ages and nationalities, who live the spirit of Ashok in their daily lives.

ENCOUNTERS

1 ***In The Beginning*** – Blowing on the Spark of Unborn Goodness

2 An Encounter with Nature's Fairytale

3 An Encounter with an Architect of Conciliation

4 An Encounter between 'Them' and 'Us'

5 An Encounter with Renewal of a Failing System

6 An Encounter with the High Priest of Financial Market Economics

7 An Encounter with a Personal Epiphany

8 An Encounter with Nature as a Partner

9 An Encounter with the Magic of Creating Shared Purpose

10 A Botched Encounter with Corporate Governance

11 An Encounter with Death from Growth

12 ***In The End*** – An Encounter with The Great Transformation

1 *In The Beginning* – Blowing on the Spark of Unborn Goodness

Two little boys were blessed by an extraordinary nanny. Eight and ten, bright and curious, they were the sons of newly mega wealthy parents, rich in money and poor in time. She was Bulgarian, brought up with the joy and initiative of a nation rediscovering itself after a long period of hibernation. She was skilled in everyday things, with an open heart which invited others, even strangers, to share and connect.

The boys loved her and sometimes called her their number two Mummy.

Every night before sleep she read or told them stories which had excited her as a child. Stories of foolish characters like Candide who believed everything in the world was the best it could possibly be, despite the most terrible sufferings. Of gallant warriors like Don Quixote who would tilt at windmills rather than backdown, if he thought they lacked virtue. Of characters in history like Gilgamesh who journeyed after truth, and the Three Musketeers who guarded their honour above all else.

One evening she read them the *Tales of Ashok* which she was reading for the first time herself. Together they found them enchanting. The children fell in love with Ashok as he went on his great journey, called by he knew not who, to sort problems which were overwhelming the world. They saw how he made connections between things which were previously neglected. They travelled with him when he soared above the world and saw things with new eyes. Above all, they loved the way he listened attentively to what people had to say and made them feel welcome.

When they had said their prayers and were settling for sleep they murmured to the nanny, "You are so much like Ashok. Can we also be like him?"

That night she cried tears of joy at the compliment they had paid her, and their discovery of the wonder of the world and the excitement of being part of it.

The next morning as they approached the bus stop on the way to school the boys took notice for the first time of the lady who stood waiting for the same bus as them every day, looking weary and worn down as though she had nothing to look forward to.

On this particular day, however, they surprised themselves by saying good morning to her in the voices of angels which children especially have. She looked startled at first but then smiled at the simple joy of being spoken to and recognised. Soon all three were chattering like birds on the wire, new friends oblivious of all the times before when they had stood silently and separately in the same place.

Nanny watched the tiny flame of Ashok kindle in the hearts of these two tiny persons she loved so much. They were born with the goodness to be teachers in the new world.

2 An Encounter with Nature's Fairytale

They were a group of rather uncommon people – artists and teachers – who gathered around Kestutis that night. They had to be, to overcome the forty years of brutally enforced suppression of public gatherings and independent opinion unsanctioned by government.

Suddenly the political lid was lifted in 1989 with the breakup of the Soviet Union, and it became possible to talk about the devastation of the forests, rivers and wildlife which had been nature's fairytale called Lithuania.

Kestutis and Dalia had a dream that by rediscovering nature as a system of interests they could help rebuild civil society in a country which had been dulled by decades of meaningless procedures and bureaucracy.

In a country where to be visible was to be vulnerable.

Together they designed a project whose actors would be children, teachers, families and communities collaborating to bring art, biology and the environment together. Its highlights were to be the works of individual children from all parts of the country, assembled as a touring exhibition.

Crucial to the project's dream was showing the child's name prominently in the catalogue, unhidden and in plain view. Families shared their children's experiences at school and gave pride of place in their homes to the catalogue, opened at the page which showed their child's contribution.

The exhibition attracted large and fervent crowds wherever it went, encouraging a public conversation about a new way of life.

The complex process of remaking a functioning society began from this group of artists and teachers who worked for no money, had a vision of change in their daily lives, and created communities that cared for each other and the environment they shared.

Kestutis' vision of nature as a system of interests that act on our feelings to create collaboration and independent commitment became more than a fairy tale of unending beauty. It created a way of being together in society based on pride in co-operative achievement and the right to be engaged.

3 An Encounter with an Architect of Conciliation

Brendan had a gift and a passion.

He had been gifted with a remarkable singing voice which put him front and centre at pub nights and social gatherings. Then he was invited by the Armagh City Choir to a rehearsal in the Cathedral, where he found a community of like-minded souls and voices who met simply to make divine music together.

His passion from childhood was to play the kettledrum in the great processions which celebrated the historic battles between Catholic and Protestant communities. Politics had resorted thirty years ago to the bullet and the gun and less publicly he had become an active member of the paramilitary wing of the Ulster Defence Regiment.

He sometimes wondered about the message in the book of Exodus, which had been a school prize many years before, which read, "You must not ill-treat an alien because you were once aliens yourselves in the land of Egypt." But it only just hovered in the margin of his life which had been entirely spent without meeting people 'of the other sort', as they say in Northern Ireland.

Then along came a man called Michael O'Donnell, who literally transformed his sense of how the world could be and who he could be in it. Michael's magic was to see that wonderful choirs existed on both sides of the border, which could be yet more wonderful together, but had no connection. So he created the Cross-Border Project which without hectoring or lecturing brought choirs from across the divide together to meet, sing and relax in each other's company. Within a few months of the Waterfront Hall opening in Belfast half a dozen choirs from across the divide were performing 'Haydn's Creation' in a jubilation of togetherness.

Music had opened the trapdoors of fear which had previously dominated their minds. Individual choirs like the Satanta Choir from Dundalk and the Armagh City Choir went on to create their own events extending the celebration of togetherness. People became works of art with shared objectives. Thanks to Michael, a bridge had been built across separateness and anchored in respect.

Brendan meanwhile continues to use his gift and his passion, but now makes music without distinction between people of 'our sort' or the 'other sort'.

4 An Encounter between 'Them' and 'Us'

The flank of Montgó Massif in Eastern Spain facing away from the sea is a joyous cultivation of parallel terraces of olive trees snaking endlessly around the contours of the mountain. It is an astonishing physical work of great beauty built by thousands of human hands. They were not ordinary hands that did this creation, they were the hands of lepers living for eternity in colonies outside civilised society and working together across the generations.

For centuries they had inspired fear of contagion. As lepers they were 'them' to be avoided at all costs, but in the early 2000s medical science found ways to greatly lessen the effects of the disease, so they could now look for work outside the colony and take part in general social activities.

They could now be part of 'us'. Except in reality, they still couldn't. Employers, teachers, shopkeepers had lifelong prejudices against them, and colony residents had no relevant skills or cultural experience.

It was then the Sasakawa-India Leprosy Foundation was formed to bridge that gap by educating people outside the colonies and training those inside. Its vision was to develop feasible livelihood streams that drew lepers away from begging for alms.

The way, developed in consultation with both communities, was to provide seed capital to start petty household level enterprises. Similar to what Mohammed Yusuf was discovering in Bangladesh, the competency and the creativity were waiting in abundance to express themselves. To meet the high demand and create accountability, under the visionary leadership of

Vineeta Shanker, grants were converted into loans to be repaid not to the Foundation but to the colony itself to be redistributed to others chosen by the community.

Some startups worked better than others. The failures succumbed to greed for individual gain. The successes were inspired by an aspiration for a collective good where lepers would be accepted and integrated as equal members of society. Key was trust and communication, built carefully and continuously nurtured. Many made the transition from begging to earning. But the real triumphs, unmeasurable and unreflected in the numbers, were the gifts of confidence, dignity and pride as 'them' became an integral part of 'us'.

5 An Encounter with Renewal of a Failing System

Healthcare systems are under enormous pressure worldwide. Demand is increasing, the pattern of disease is changing, costs rise inexorably, morale of doctors and nurses is low.

In response governments demand greater efficiency, 'do more with less'.

That will not solve the deep-seated problem of a system designed for one set of circumstances which no longer tell the whole story. Or address the whole person. Or find out what people really want from their health service and how they want it delivered.

One alternative story gives hope. It starts in the 1980s in Alaska. The settlement that ceded the state to the USA, guaranteed the native community free healthcare. By the 1990s the system clearly wasn't working. The quality of the service delivered mirrored the state of failing health systems elsewhere in the world. Waiting times, patient frustration, staff disillusionment, health performance measures were all amongst the worst in the USA.

Things got so bad that the Alaskan Native Community negotiated to take over the system and run it themselves. The government agreed on condition there would be no increase in funding. South Central Foundation, one of a number of native healthcare providers, under the inspired leadership of Katherine Gottlieb, undertook a two year community listening process to find out what people really wanted from their health service and how they wanted to be treated. Then they delivered it.

The result is a system where families and communities have taken ownership of their own wellness and recovery. The model is described by the native word *Nuka* meaning a 'strong, large, living structure'. Patients are called customer-owners and take an active role in decisions about their own health and how care is provided. The approach integrates facility providers, clinicians, staff, patients and the community. Accountability is shared for achieving the vision of physical, mental, emotional and spiritual wellness.

Performance measures have been exceptional and rewarded with the highest level of national recognition that a US public service organisation can receive. *Nuka* has been described as the future for US healthcare and healthcare around the world. It is an astonishing story.

6 An Encounter with the High Priest of Financial Market Economics

The young man at my side looked uncomfortable. It was a dinner at a famous School of Economics in London for a distinguished Nobel laureate from Chicago. He had asked a question of the great man which did not find favour.

During his lecture the laureate had argued that the only way forward for capitalism and the free world was to radically reduce the burdens of social obligation placed on business. All business, but especially venture capital which created jobs and profits. "Let white horses run free," was his phrase, "and watch productivity zoom."

Now he had turned on my colleague and asked in a way which indicated anyone who disagreed was an idiot, "Surely you believe in innovation? It's what drives growth. Everything else is rent from dead capital or bureaucracy".

The dinner was taking place in the aftermath of the great financial crash of 2008. The young man had ventured the thought that it was animal spirits running free from regulation that had destroyed jobs, homes, consumer confidence and the very growth he was advocating. Worse it had been disastrous for planetary health.

The silence that followed was one of those when even brave souls sit on their hands. The young man had been hauled in at the last moment to make up the numbers. Now he was being upbraided by a laureate with enough academic badges to sink a flotilla of junior lecturers, who promptly proceeded to do just that. One by one he unpeeled the layers of lost productivity equal, he said, to one third of annual growth. When he put the alternative case that measures of GDP and earnings per quarter were not only outdated but dangerous, witness ecological collapse, levels of sickness, and absenteeism, the young man was shouted down. His more senior colleagues were equally intimidated, possibly wary about their careers or just exquisitely courteous hosts.

I learned later that young man had gone on to create a planetary dashboard for resilience in land and watershed systems and growth in pollution-based illnesses at the nearby School of the Common Good. In his spare time he counsels the growing number of families experiencing death and illness from preventable environmental diseases. He admits he still burns at the thought of the encounter with the high priest, but believes it was also the point at which he was able finally to turn away from market economics as infallible dogma.

7 An Encounter with a Personal Epiphany

Where do they come from these revelations that change people's lives so completely they seem to have been born again? A new person who looks like the old one and has the same name but thinks differently, feels differently and is obviously committed to a new way of being.

Like Ashok, I heard a voice a long time ago calling ever so faintly. I was half-awake watching the first light of day touch the Atlas mountains brooding, all powerful, behind the orange groves which led the near eye from the oasis of beauty called the Gazelle d'Or deep in the Sahara to a snow-capped distant city in the sky.

Unlike Ashok, I didn't know what to do with it at the time, and went back to a perfectly pleasant life as a gadfly, pecking at the carcasses of businesses that had lost their way, adding a shoulder here and there to organisations interested in the arts of renewal, trying vainly for the most part to add something to an academic institution dedicated rather wonderfully to 'understanding the causes of things'.

It was only much later I would meet the man and his book which would complete the epiphany. As a young academic labouring in the foothills of academe, Maurice Glasman had written a book called *Unnecessary Suffering*. For me it was a bolt from the blue. I could suddenly see clearly the other side of the world of wonder, discovery and creativity we are fortunate to experience in good times, glimpses or prolonged exposure if we are lucky, depending on where we find ourselves in our personal journeys.

The reverse side was one of Suffering, part of which is unavoidable and a large part of which is Unnecessary. Death, uncertainty, illness, loss of loved ones, the slow decline from consciousness to confusion and forgetfulness are all part of the human condition. But a large part of our suffering is of our own making. Aggression that pits one nation against another, hunger so deep that bodies ache for release, inequality and conflict and environmental destruction so extreme that millions of people displace themselves thousands of miles in the belief there must be a better place to be.

It was the recognition of that immensity of unnecessary suffering that drew me to the pursuit of the Common Good and to join Ashok on his journey towards finding a way to a better and more noble place.

8 An Encounter with Nature as a Partner

There is a river in Wales which snakes between low hills and on certain nights glows like liquid silver in the light of the moon. In daylight, it can be so still that it's only by the gurgles it makes as it slides across fossil-filled stones from another age that you know it's alive.

A man from London bought it and fell in love with its independence and connectivity with places far away. But this was ownership of a kind he had never encountered before.

First the river was alive and mobile, changing its pace and dimensions without any reference to him. The winter rise could be ten feet or more, the narrow ribbon becoming a moving mass of water as wide as a motorway. Overnight it could freeze into a solid slab of ice, roofing over fields and hedgerows.

But it was also a prisoner. It had no control over the effluent which ran off fields further upstream nor the toxic waste from the newly built battery hen farm over the hill. It had no say in the destruction of its protective hedgerows and habitat as the buckets of giant bulldozers left the wounded limbs of oaks and sallow hanging in the wind. Kingfisher, heron, salmon and otter disappeared with the trees in a re-run of Gilgamesh taking his axe 4000 years ago to the cedar forests on the banks of the Euphrates. Now, very late in the day, river 'ghosts' are being resurrected and river 'rights' accorded in dozens of countries since the first epic decision in New Zealand to declare the Wanganui River alive with the full rights of a legal person.

These ideas have now washed up on the shores of other countries. In Britain charters of rights are being drawn up for the Ouse, the Don, the Derwent and the Clyde. A group of Guardians protects the River Roding in Essex. In 2021 Canada recognised the rights of the Mutuhekau Shipu not to be drowned by a hydro dam project. In the same year the cloud forest Los Cedros was accorded the legal right to life under the Ecuadorian Constitution.

It might be a step too far to similarly honour the beauty of that liquid silver tributary of the Wye in Wales, but ecocide of our rivers, by design or default from mining, pollution and damming is increasingly seen as contrary to the 'natural contract' between man and water. Gilgamesh is at last sheathing his axe.

9 An Encounter with the Magic of Creating Shared Purpose

It wasn't from a mountaintop or big data that the problem became clear. It was from close up contact with the very poor in their homes and villages. Particularly with the women who made a more significant difference to the families' welfare than the men.

He noticed that when they could get credit, they made craft products out of bamboo and grew vegetables which they sold at a profit in the market. There were two problems holding them back. Commercial loan rates were very high and the women were frequently robbed of their earnings on the way back from the market.

His solution was to provide them with very basic mobile phones for their security and to advance them tiny amounts of credit under his own guarantee. Groups of five worked together and assumed shared accountability for repayment. The recovery rate and the reinvestment of savings were far higher than any other bank. Grameen Bank was born and now operates in 94% of Bangladesh's villages. The model has been replicated in 64 countries, servicing 300 million people.

The man, of course, was Muhammad Yunus and he went on to focus on one other area of massive strategic importance. Bangladesh is a low-lying country prone to flooding. As a result of widespread poverty and disease, the country ranked very low on the Human Development Index. People resisted vaccines, but liked yoghurt. Making the connection Yunus persuaded Groupe Danone to partner with Grameen in the production of Shokti Doi, a cheap yoghurt containing 30% of the micronutrient's needed for a child's development. The results in terms of improved health have been astonishing.

It is a wonderful example of business and political community bringing together technology and local distribution skills in a 50/50 joint venture which pays itself no dividends and reinvests all profits. It took two remarkable men, Yunus himself and Franck Riboud at Danone, to start from apparently insuperable modern-day problems and work backwards together to deliver shared solutions, straight from Ashok's playbook 5,000 years ago.

10 A Botched Encounter with Corporate Governance

Companies create wealth, provide jobs and shape the values and behaviours of employees, customers, suppliers and the community at large. Their governance is therefore supremely important but the system we have is destructive of national wealth, integrity and meaningful personal purpose.

Take integrity. A generation ago a very senior banker was found by the Presiding Judge to have been 'less than frank' in his evidence. The culture at Warburg was that behaviour that detracted from the firm's integrity was not acceptable. His resignation was immediately accepted. The message was clear. Accountability is personal and 'right' behaviour is intrinsic not procedural.

Since then corporate governance has become one of the fastest growing corporate activities. Numerous codes have set down rules for Board structures and composition and procedures at every level of customer interface and employee engagement. Governance has become compliance. Lawyers and Human Resources have become the gatekeepers of integrity.

But wrong things keep happening. Enron, the Police, Grenfell Tower, miss-selling, environmental destruction. Rarely were they the result of someone deliberately choosing to do wrong. What was 'wrong' was the way we decide what is 'right'. Control has replaced trust but the mesh of compliance regulation can never be fine enough to capture the risk complexity of a business. Boards get misleading assurance from their audit and compliance functions. Worse they approve incentive schemes which enshrine risk and reduce personal accountability.

Take productivity. When the workforce is defined as a commodity not as the core competitive asset of the business, it is unlikely to 'own' the longterm performance of the business. Reward ratios of 150 between the top and bottom create resentment not commitment. If the duty of governance is to embrace the interests of shareholders, the development of its people, and an inalienable obligation to do no harm, this is not the way to do it.

The work led by Paul Polman is foundational for an industry-wide epiphany. The decision by the Grant family to hand over ownership of The Entertainer to employee trusts is a practical demonstration of placing 'right' values at the centre of governance. Governance conceived more widely in this way is an encounter still waiting to happen.

11 An Encounter with Death from Growth

They gathered around him as the afternoon shadows grew longer. The fields surrounding them were peaceful, mantled in a light mist. In contrast, just over the brow of the hill, the air was thick with the smoke of forest fires which had been burning for days. In the valley behind a torrent of mud had swept houses and drowning livestock tumbling towards the sea. Out of sight the 500,000th victim in Ukraine sank to the ground, sightless and limbless, targeted by a drone now long gone.

So far, their hillside had been a refuge from the violence that was going on around them. But they had become increasingly gripped by fear knowing their isolation couldn't last. Within the community violent confrontations were already breaking out as people competed for what was left of the drinking water at the well and took possession by force of the most strongly built houses for their own protection.

As the situation got worse, for many days and nights their lamentations had filled the air as they asked themselves time and again why the Gods had disowned them. They found no answers. Some raved to the heavens or lapsed into a stupor as they lost their minds. Others decided to help themselves, seizing whatever they could for themselves. Someone finally suggested in desperation they should go to the wise man on the hill and ask for guidance.

Now as they sat together on the hillside, they spoke to him of their fear that the surrounding sea of fire, flood, and disease would overwhelm them. Or be laid waste by violent newcomers displaced by others in their turn. The community of settled relationships they had enjoyed for so long was breaking apart before their eyes and they didn't know what to do to prevent it.

Strange, they said, that they had learned how to impose their dominance on the fauna and flora of the natural world and to create in their brief existence alongside that dominion a world of wondrous sensitivity and artistic richness. Now that was all being extinguished. They were on a bus heading for collective disaster which was very close to its destination and they wanted above anything to stop the bus and be allowed to get off.

They were confused at first when the wise man asked which of them on their glorious journey to this point had decided to acidify and overfish the oceans, pollute the streams and rivers, and mine the innermost depths of the earth. Was it the ancient Gods, or had they unwittingly substituted a new God in their place? If they had, perhaps it was in the nature of that God that they should seek answers to their prayers.

They fell into confusion and resisted angrily the idea they had been party to the environmental crisis they were now so desperate to avoid, including the culture of commercialisation of everything that could be turned into profit by intense competition or military aggression. When the anger was finally spent, someone asked whether it was just a coincidence that, when wars and covid put growth on pause, the oceans were replenished and clearer skies and the bravery night after night of the previously unconsidered emergency services opened a new kind of social and spiritual imagination?

Slowly the idea developed in their minds that, dreadful as those events were, they forced into their thinking the possibility of a new way of being together in society. Since our unique strength as a species is the ability to think and make choices, the realisation that they themselves had brought into existence the world that was threatening to engulf them, by their thinking and their decisions, ran like a shudder through the crowd.

And then just as quickly the shudder turned to excitement as the idea took flame that, if they brought that world into existence by their thinking, it followed they could build a new world by thinking anew. It was suddenly clear that if 'I am only for myself' and the new fundamentalism of material growth and consumption is my God, the world we will have in the future will still be the world we are now fleeing. And we would be right to be fearful.

By the time night had settled over the encounter, the shape of an alternative way of growth had begun to form of purposeful co-operation for shared goods of health, security of employment, education and partnership with the environment, alongside individual wellbeing. As they turned to go down the hillside, they felt the strength of a new shared purpose beginning to displace the fear and uncertainty they had brought up with them.

They thanked the wise man for his willingness to listen, and took with them in their souls the stirring of an understanding that the duty of those who have the privilege of encountering Ashok was to go forth as teachers and create encounters of their own, opening new ways of thinking.

12 *In The End* – An Encounter with The Great Transformation

The present times are confused and conflicted. Wars, migration, gross inequality and climate change are signs of a sickness in society which is embedded in our way of thinking which is linear and exclusive, our priority of material growth, our way of organising which is national and partial, and our values of competing and winning.

They are not the solution to our confusion, they are its cause and preclude the great transformation which must take place to preserve the grace and wonder of the human race.

These encounters describe some of the wonderful sparks already happening – which could ignite that profound transformation, if they were fanned by the breath of a new enlightenment to go forth and multiply. How might that happen?

Laurens van der Post, who embodies the spirit of Ashok, tells the story of living with the bushmen in the Kalahari and learning their ways. In the long drought before the rains come, like all living things in the desert, they wait with desperate anxiety for the lightening to come, the signal of a new season of plenty. When it did burst after a long period of darkness, it was like God beckoned them and, having rediscovered their sense of purpose, they had to move towards the light. They were instantly transformed.

He was transfixed by two events. Against the water light of the stars emerged the outline of a woman holding a child high above her head, singing softly with her face lifted towards the sky. She was asking the stars to take the little heart of her child and to give him something of the heart of a star in return.

Then an old man talked of the magic of the steenbok. Too beautiful to be aware of imperfection, too innocent to know fear, too gentle to suspect violence. Disturbed, it would stand quietly by its 'place' and look without fear out of its great eyes to see what the wonderful noise could be about. The magic was that the hunter began to feel there was a steenbok person standing behind him willing him to act in the same way.

The Great Transformation will occur for us when we too invite the stars into our hearts and feel the presence of a steenbok person in our souls, guiding our values and our behaviour in the way of Ashok.